PREVIEW COPY
Compliments of
STEVE DAVIS PUBLISHING

THE ELECTRIC MAILBOX

Steve Davis
Candy Travis

ST. PHILIP'S COLLEGE LIBRARY

Copyright © 1986 by Steve Davis Publishing

All rights reserved.

Steve Davis Publishing
P.O. Box 190831
Dallas, Texas 75219

ISBN 0-911061-14-2

Library of Congress Catalog Number 84-91756

First edition.

10 9 8 7 6 5 4 3 2 1

CONTENTS

	Introduction	3
1.	Master of Your Own Communications	5
2.	Making the Connection	24
3.	AT&T and Other Voices	58
4.	The Source	63
5.	CompuServe EasyPlex	80
6.	Delphi	93
7.	ECHO	109
8.	Dialmail	119
9.	MCI Mail	131
10.	EasyLink	148
11.	InfoPlex	166
12.	GTE Telemail	179
13.	RCA Globcom	199
14.	OnTyme	210
15.	GE Quik-Comm	224
16.	ITT Dialcom	233
	Appendices, Network Access Numbers	253
	Index	282

About the Authors

Steve Davis has been writing and publishing books on computer-related topics for over four years. Electronic mail has played a vital role in every project, providing a quick, cost-effective communications link with authors, editors, typesetters, vendors and customers.

Candy Travis is supervisor for the technical writing department of a major municipal government agency. She is the author of *The Handbook Handbook*.

Introduction

If you were to stop the average person on the street and mention the term "electronic mail," what kind of reaction would you expect? Most likely, they would stare at you in stunned silence, or in awe. The term conjures up visions of Buck Rogers, or Orwellian big-government super-computers, or some sort of fourth-dimension that only the most anti-social techie could comprehend.

Welcome to the future. Electronic mail is not some futuristic fantasy, but it is part of your future. However, you don't have to wait until 2001 to begin enjoying the benefits of this efficient form of communications. You can join this "quiet revolution" in communications *now*. This book will be your guide.

The Electric Mailbox has been designed to introduce you to a variety of popular electronic mail services that are available today, to help you select the service that best fits your individual needs, and then to get you up and running with minimal effort.

The first chapter, "Master of Your Own Communications," provides an overview of the world of electronic mail communications and sets forth general guidelines for choosing a service. The second chapter, "Making the Connection," tells you what you need in terms of hardware and software to get started. In the third chapter, "AT&T and Other Voices," you will be introduced to a relatively new breed of electronic mail called "voice mail."

Chapters Four through Sixteen fulfill the promise offered by the subtitle of this book, "A User's Guide to Electronic Mail Services." Each service is described in terms of cost and capability. But, more importantly, you will learn how to *use* the services to gain the maximum benefit. Each chapter

The Electric Mailbox

contains step-by-step instructions for sending and receiving mail using the services described. Simple lessons will show you how to:

- Access the network.
- Logon to the service.
- Get help.
- Send a letter.
- Upload and download files.
- Create and use mailing lists.
- Work with online files.
- Read a letter.
- Send telex messages, telegrams and paper mail.
- Logoff the system, and more.

By comparing the features and ease of use of several systems, you will be able to make an informed decision as to which service (or services) will fit your needs.

Finally, for your convenience, the appendices contain local access telephone numbers for each of the major communications networks.

When the Pony Express "revolutionized" message delivery, speeding urgent mail from St. Joseph, Missouri to Sacramento, California in 10 days for only $5.00 an ounce, who could have imagined that someday it would be possible to send those same time-critical messages anywhere in the world in a matter of seconds for just pennies?

That day is today. The world is at your fingertips. So what are you waiting for? Let *The Electric Mailbox* plug you into the exciting new world of instant mail.

Master of Your Own Communications

Never again will you be harassed by ringing phones, costrained by divergent time zones or victimized by postal delays and strikes. From now on, you will be master of your own communications with all those you deal with professionally. This is the potential of electronic mail.

Martin Lasden, *Computer Decisions*

You know you are in trouble when you hear the bored voice of Miss Jones again. "Oh, Mr. Smith," she says. "Mr. Brown got your message from this afternoon when he got back from his board meeting and he called you right away. But your secretary said you were out to lunch. It's 5:01 here now, and of course Mr. Brown has gone home. Would you like to leave a message again?"

If and when you ever get to talk to Mr. Brown, how much of your valuable time is wasted on small talk, and how many times is your call interrupted? "I've got better things to do," you mutter to yourself, trying to endure a saccharine chorus of violins while on hold.

The only winners in this game of "telephone tag" are the long distance carriers and the manufacturers of those little pink "While You Were Out" pads. You and your business are the losers.

And, of course, you don't even get to play telephone tag if you are in New York and the person you need to talk to is clear around the globe. At noon in the Big Apple, it's 3 a.m. in Sydney; as New York closes down at 5 p.m. on Friday, they're sleeping in Down Under because it's 8:00 on Saturday morning. You could try to send a telex to your agent in Sydney, but the last clanking teletype was taken out of your office almost 20 years ago.

The Electric Mailbox

And then there is the dreaded mailroom, the place where important letters and packages seem to go to die. Many an efficient, high-powered company has been brought to its knees because of the incomprehensible bottleneck of its mailroom. Interoffice mail from a desk in one corner of the executive office destined to a desk at the other corner would travel downstairs to the basement for sorting and distribution. With luck, a message would make it across the room in a few days.

Important incoming mail from vendors and buyers would also travel through the maze of the mailroom, emerging on a schedule guaranteed to enhance ulcers. How many times has an urgent Federal Express package made it all the way across the country in just a few hours only to languish for days under a stack of week-old "Wall Street Journals" in the mailroom?

And finally, there is the joy of working with messenger services, companies which seem to take great pride in staffing themselves with illiterate kids with no sense of responsibility or direction, all the while charging outrageous fees for their service.

We are not just talking about mere convenience here. The fact is that the pace of national and international business has increased tremendously in the past few years as a result of the proliferation of microcomputers and telecommunications interlinks between them. Stock and other financial marketplaces are now open for trade virtually fulltime, and customers now expect nearly instantaneous service on their orders and queries. Perhaps most importantly, your boss, who may have a personal computer on his desk and the ability to perform his own research and report-making, expects your contributions immediately.

The solution to all of these problems of modern-day business is a surprisingly simple update of the old, reliable (if slow) postal service: electronic mail. And that's what this book is all about. In this book, you'll learn how to make use of this new, electronic letter carrier in your business and personal operations. You'll find out how to equip your office for telecommunications, how to choose an electronic mail service, and how to get the most efficient, economical use from the carrier you have selected.

What's So Wonderful About Electronic Mail?

Computers and telecommunications and electronic mail are all sexy words in today's world of high-tech, but as a business person it still comes down to the bottom line: What is so wonderful about electronic mail? The answer is encompassed in a single, powerful word: *productivity*. Here's some "for-instances."

- Your company can maintain regular contact with customers all around the world, in "writing," 24 hours a day, 7 days a week. Your customers can be subscribers to the same electronic mail service, or another service with a gateway, or telex users. Company executives can check the "mailbox" at any time, too. Many electronic mail services allow users to use pre-defined forms for orders and inquiries.

- Your sales people on the road can place orders and communicate with the shipping department and the factory from motel rooms and restaurants and the offices of customers.

- Your executives, traveling around the country, can pick up their memos and send off answers and new correspondence from wherever they are.

- Your far-flung sales offices and manufacturing sites can receive instant copies of corporate memos and announcements.

- Your company's consultants and contacts can maintain a two-way information pipeline.

- Your company can conduct negotiations and perform research with specialists all around the world, regardless of location, time zones, and working schedules. In fact, chapters of this book flew back and forth between Texas, New York, San Francisco and elsewhere, using the facilities of an electronic mail service.

The true bottom line, then, is this: Electronic mail is not only a savings in time and expense as a replacement for certain types of existing communications, but it is also a doorway into new areas of business.

Echoes of the Pony Express

One of the advantages of using the postal service for sending a letter is that you, the sender, can compose your letter at your own convenience and post it as soon as it is done. You do not need to coordinate your activities with the postal service or with the person to whom you are sending the mail. Similarly, that person can go about his or her work without paying any more attention to your activities than to check the incoming electronic mailbox once a day. A major disadvantage, of course, is that the postal service may pick up and deliver your message at *their* convenience, not yours.

The Electric Mailbox

Though mail handling has become increasingly automated in recent years, the postal service is still a highly labor-intensive operation, depending upon workers to pick up mail, sort the letters at several post offices or bulk mail centers, and then deliver it to the addressee. (It has been estimated that over 80% of mail costs are labor-related, while only 7% relate to the actual transportation of mail.) Although we may all regularly complain that a letter may take a week to travel across the country (or across town), some might argue that any other assumption is an unreasonable one for the price we pay.

So, too, it is with so-called "express" delivery services. You can take a letter and put it into an envelope and call Federal Express, or Emory, or Purolator, or any of dozens of such companies, and have the package picked up and delivered "overnight." The time lag on the speediest service can range from as little as 13 hours (pick up at 8 p.m. and delivery the next morning at 9 a.m.) to more than 60 hours if you miss deadlines (a pickup at 8:01 p.m. on Friday might not be delivered until Monday morning.) And such services are relatively expensive; a small envelope typically costs about $11 to send by express mail, with larger shipments costing more. If the material is in electronic form, the user must go to the trouble of printing out a copy or send a floppy disk (in a compatible format) in the express pouch.

Another alternative is the use of a facsimile transmitter or service. If your company has its own system, you'll have to print out a clean copy of the message and then feed it into the facsimile transmitter. At the receiving end, there must be a compatible machine. Once the message is sent, it must be delivered to the addressee. Facsimile is faster than courier services, although nowhere near as quick as electronic mail. As facsimile units become integrated with other office communications equipment, such as personal computers, it may become an increasingly popular method of transmitting paper documents quickly, especially those containing graphics and signatures. However, facsimile has two major disadvantages: one is that material being sent and received is not necessarily hidden from the view of unauthorized eyes; the second is that you can't send a file in electronic form using a facsimile system. If you need to put received copy back into electronic form for editing and re-transmission, you'll have to have it retyped or fed through an optical scanner. With either system there is lessened security and the possibility for the introduction of errors.

Electronic mail, on the other hand, is a highly computer-intensive operation. It essentially follows, though, the same pattern as the postal service. It uses a computer to pick up your incoming mail, sorts messages by computer, transports the mail electronically to your recipient's "mailbox," and then delivers the message when the addressee asks for his mail.

Electronic mail grew from humble beginnings: workers at terminals connected to huge mainframe computers in the hoary days of the 1960s discovered they could send little messages back and forth from one screen to another for as long as they both were signed on. At that point, the system was little more than a typewritten equivalent of a telephone link.

It was the development of the concept of a "store and forward" system that made the interconnection of computer screens such a valuable tool for business communication. The concept is simple: instead of having the computer merely serve as a pipeline from one terminal to another, an electronic memory bank is installed between User A and User B. In this way, a message from your terminal is delivered to the "post office" and kept on file there with a forwarding and return address. Your recipient signs onto the computer and asks the post office to deliver the mail addressed to him, and the computer unloads the contents of his mailbox along the communications line.

Now, Mr. Smith can compose a message and send it on to the memory of the main computer, which delivers it instantly to Mr. Brown's mailbox. Later, when Mr. Brown signs on, he can retrieve that message to his own screen or printer. And Miss Jones can tend to her filing, or her nails.

What is Electronic Mail?

Electronic mail, or "email" as it is called by converts, is more than just the ability to send short messages from one computer to another. It is, in fact, an electronic equivalent of nearly all of the important business services offered by the postal service. Whatever is in your computer can be sent to just about anywhere in the world, almost instantly in electronic form, and very quickly in printed form.

In this book, you'll learn about a whole range of special services offered by the various email companies. Some of these include facilities that allow you to:

- Send an electronic file from your desktop computer to a high-quality laser printing station in a major metropolitan area near the addressee. There the letter or file is printed out (you can even have the system add a computer-stored facsimile of your letterhead and signature) and then put into the US Postal Service mails for local delivery.

The Electric Mailbox

- Send an electronic file to a laser printing station and then have the resulting printout hand-delivered to the addressee in just a few hours.

- Send or receive domestic or international telex messages to and from over 1.6 million teletypes around the world from your desktop computer.

- Send an electronic file to a central computer where it can be converted into artificial speech and "read" to a designated recipient over the telephone.

Special services associated with some electronic mail services include the ability to request "receipts" of delivery when mail is accepted, COD or toll-free mailboxes, and pre-addressed reply notes to incoming mail. You can also automatically send "carbon copies" or "blind carbon copies" of letters to others, and you can create your own lists for automatic mass mailings.

Today, you can send and receive electronic mail in one of three ways:

1. By signing up as a subscriber to one of the national electronic mail utilities, such as MCI Mail, EasyLink, CompuServe, or The Source.

2. By contracting with a national telecommunications networks such as Tymnet or Telenet for a specially designed mail service for your company.

3. Or by setting up your own service, based on your company's own computer facilities or those of a timesharing service.

The principal advantage of using a national public network is the access to users outside of your company. If you end up with a private service, you might want to investigate addition of a gateway to one or more of the public access services so that you can exchange messages with users on these systems.

How to Choose an Electronic Mail Service

You'll find in this book descriptions of the wide range of services offered by many of the major electronic mail providers. Read them carefully to find the services you require.

Here, though, are some general guidelines to keep in mind as you begin the search for the best service to fit your needs:

- *Access*. Are the people or companies that you want to communicate with already on the same network? The best service in the world is of no use to you if you have no one to talk to; it is the electronic equivalent of one hand clapping. Are there gateways to other networks that expand your reach? Are there alternate means of sending messages, such as connections to the national and international telex systems, the US Postal Service, or to courier services?

- *Security*. Does the electronic mail provider offer a level of security appropriate to the nature of the messages you will be sending and storing? Some simple bulletin board services include electronic mail, but messages are open for all to read. Are passwords required before mail is revealed? Some services require special passwords for individual messages as well.

- *Batch services*. Does the electronic mail provider offer users the ability to send one message to a mailing list of recipients that is either kept on file in the service's computer or is submitted by the user from disk or keyboard?

- *Error-checking protocols*. Advanced services have begun to implement one or more error-checking systems designed to electronically "proofread" your files. Such protocols include the Xmodem and X.PC systems. If your electronic mail messages contain critical details, such as financial figures or prices, an error-checking scheme that works with your telecommunications software can save you a great deal of trouble and embarrassment.

- *Text editors*. It is always easier, and usually cheaper, to compose and edit a text message using a word processing package on your computer and then "upload" it to the electronic mail service for delivery. However, there are times when an online text editor can be of help, such as when you are away from your office and using a portable terminal without a text editor, or if you are responding quickly to a message you have received. Is there an extra charge for using the system's text editor? How difficult is it to edit copy? Will it format text to appear in a business-like form?

- *Binary and special files*. If you will be needing to send programs or other special files, such as spreadsheets and database files, you'll need to know how flexible the network is. Will the service allow you to transmit program files in a binary form? How about text files composed using a word processor that uses special non-text characters for formatting? Can you attach a binary file to a text file and send them both at the same time?

- *Special services.* How important are post office-like services such as return receipts for letters delivered, and electronic forwarding of messages? Does the service offer these options, and is there an extra charge for them?

- *Network access.* How easy (and inexpensive) is it to gain access to the electronic mail service? The best of all worlds is a service with free or low-cost local access in major metropolitan areas, and a nationwide toll-free number for other areas. Be aware, though, that there is no such thing as a free lunch or a free link. You'll pay for access in one way or another, either on your own telephone bill or in charges levied by the electronic mail provider. Compare the total cost of any service, including these factors.

Hard Facts About Hardware

The amazing thing to most electronic mail users is how very simple the whole process is to begin. The advent of the desktop microcomputer and the tremendous expansion of telecommunications it has brought has resulted in at least some standardization. Here's all you will need to begin using electronic mail:

- Any standard business or professional personal computer, such as an IBM PC, Apple II or Macintosh machine. Devices from companies including Radio Shack, Xerox, Kaypro, Compaq, and others will also work. Even most inexpensive home computers, such as the TI 99/4A or the Commodore 64, and "dumb" and "smart" terminals, such as those made by Qume, Hazeltine and others, have communications capability and may be used for sending and receiving electronic mail.

- A standard telecommunications software package matched to the computer you will be using.

- A "serial" communications connection from your computer.

- A standard modem device to convert a computer's digital data into a form that can be transmitted over a telephone line.

- A standard telephone line.

- An account with an electronic mail service.

You'll find definitions of all of these terms, and a guide to purchasing hardware and software, in the chapters that follow.

Toting a Portable Load

Not all telecommunications devices must be rooted to the desktop. The microcomputer itself was a remarkable advance in miniaturization, shrinking the power and capability of a huge room-sized mainframe computer to a small box that could sit on a desk. The incredible shrinking computer did not stop there, though. Portable, or "laptop" computers now allow professionals to do much of their work on the road...in hotel rooms, on board airplanes, or at the offices of clients. These devices can then be linked back into the electronic mail network with the use of a portable modem.

Sales representatives out on the road can easily enter their orders; insurance salespeople and adjusters can consult their in-boxes and file reports from the field; accountants and CPAs can visit clients to perform their work and upload and download data.

The design and operation of a portable computer is not much different from that of a full-sized microcomputer. With present technology, it is possible to shrink a fully capable microcomputer down to the size of a paperback book or even smaller. The limiting factors are ergonomic: How small can a keyboard be and still be useable? How small can a display screen be and still be readable?

The first portable computers were more properly called "transportables," and then only for those with strong backs. A typical unit weighed about 35 pounds or so and was packed into a case that was (as they used to say on "What's My Line") bigger than a breadbox.

The new technologies that made the true portables possible were the development of battery-powered electronic parts, and the liquid crystal display. Typical LCD screens today display at least 8 lines of 80 characters. Some units offer "full screen" displays of 25 lines by 80 characters, the same as most video screens. However, bigger is not always better, since the larger LCD units tend to suffer from fuzziness and distortion caused by glare. Thus, the LCD screen is both a blessing and a curse; the letters on its face are readable, but many users will find it difficult to view. You'll have to make your own, subjective decision.

Many portable units now include a built-in modem. Virtually all portables also include a serial communications port which can be linked to an external modem. One of the advantages of using a telecommunications link

The Electric Mailbox

from a portable computer is that it removes any need for "compatibility" between a portable unit and a desktop or other computer elsewhere. Files transmitted to and from an electronic mail service are usually pure and simple data and not dependent upon a particular computer operating system or software program.

Here are some questions to ask as you set out to purchase a portable computer for use in an electronic mail system:

- *Do you need disk drives?* Permanent storage devices are available for many portable units, but they take up space, add weight, and cost several hundred dollars each. For many users, the battery-powered RAM (random access memory) storage of the typical portable unit will suffice for the time between uploading and downloading into larger desktop units. A halfway measure is to purchase a unit that includes a built-in microcassette recorder that allows backups of the RAM for temporary storage.

- *Do you need a built-in modem?* Or, can you work with a portable modem or a fixed modem installed at regular transmission points such as your home and office?

- *How fast a modem do you require?* For occasional transmission and receipt of electronic mail messages, a low-cost 300-baud modem should suffice. If, on the other hand, you will be sending and receiving a great deal of mail, or lengthy files, a faster 1200-baud modem will save you time and connect-time charges.

- *Does the system have enough internal storage capacity?* Figure out how much material you will need to store at any one time. A "64K RAM" computer sounds large, but if it has to give over 48K to the operating system and word processing program, you are left with only 18K, or about 3000 words (six single-spaced pages of text). A good feature is the ability to add extra storage in the form of plug-in cartridges or "RAM disks."

- *Is the unit well constructed?* Remember that the device is going to be crammed into suitcases, squeezed under airline seats, and bounced in the subway. Will it stand up to the task?

- *What built-in programs are included?* Many portables now come with word processing packages, spreadsheets, and communications applications stored in ROM (read only memory). Are these ROM chips changeable? Do they include programs you want to use? Are the programs themselves capable and easy to use?

- *What is the battery life?* A unit that promises only four or six hours on a set of batteries may not be powerful enough for a long business trip. Are the batteries rechargeable? If not, does the unit accept standard and inexpensive batteries, or does it require special devices that may not be readily available away from your home base. If you will be travelling outside of the US, you may want to consider whether the unit can be recharged from foreign power supplies.

How much does the portable, fully equipped with the options you have selected, weigh? You may think that 10 pounds is a light weight at 7:30 a.m., but by 5:45 p.m., it may seem an oppressive burden in your briefcase.

You'll find capable portable and laptop units available from a wide range of computer companies, including IBM, Data General, Hewlett-Packard, Texas Instruments, Grid, Radio Shack, NEC, Epson, Compaq, Kaypro, Sharp, Sord, and Zenith.

Or Do You Want a Telecomputer?

Insisting that it was not done just for the purpose of confusing the situation, several computer manufacturers have added two new categories of telecommunications equipment: telecomputers and computer telephones. The distinctions between these two types of machines, and indeed between either of them and a regular personal computer with a modem, can sometimes be quite minor.

A telecomputer (or telephone computer) is a single unit that combines a computer and an intelligent telephone. The computers themselves are fully capable of running most applications software, from telecommunications programs to word processors, spreadsheets, and databases.

On the other hand, a computer telephone is essentially a telephone with a computer hung on the side. These devices use the power of the microprocessor as an aid to the carrying and acceptance of electronic messages, both data and voice.

Among the first telecomputers were a family of machines from TeleCompaq, a division of Compaq Computers. One distinctive element of these machines is their inclusion of no less than three floppy disk drives (one of the drives and 128K bytes of memory are dedicated to the purposes of the telephone applications software.) Like other Compaq models, the TeleCompaq machines are compatible with most software and hardware manufactured for the IBM PC series.

Various models of the TeleCompaq include a hard disk drive, from two to five incoming telephone lines, built-in Hayes 1200 baud modems, and a telephone handset with speakerphone. Specialized software includes telephone management, communications, and desktop management applications, with functions including a phone directory, electronic mail, calendar, notepad, and calculator. The TeleCompaq can carry on simultaneous voice and data communications without interrupting work in other areas.

Another IBM-compatible is the Zaisan ES.3. The keyboard has two tiers of keys. The lower is a computer keyboard; the upper is a telephone keypad with 13 programmable function keys and eight dynamic "soft keys" for menu selection. The computer comes with software that includes a telephone log and notebook, phone directory, and appointment calendar.

AT&T offers a multiuser machine called the Unix PC that comes complete with three modular phone connections, built-in modem, speakerphone, and telephone management and communications software. Features include one-key dialing, automatic last-number redial, phone directory and a log. The user can simultaneously talk on the phone, transmit a file, and view information from another file on the screen.

Rolm, now owned by IBM, offers several machines, including the Cedar computer, an IBM PC compatible machine which can accept from one to four incoming phone lines. It includes a telephone with handset, speakerphone, and telephone-management, communications, and desktop management software. A "Do Not Disturb" command halts or redirects all incoming calls. "Executive Override" lets you interrupt a call on another extension within a Rolm PBX system. Users can also conference with as many as eight phones, two of which can be outside of the PBX.

Moving on to computer telephones, one of the first such machines is the Cosystem from Cygnet Communications. The Cosystem is a console with buttons and a telephone handset, with a cable running to an internal adapter card mounted in an IBM or compatible computer. The Cosystem can communicate with any other standard modem or communications service, but it is particularly powerful when connected with another Cosystem at the end of the phone line. The user can send messages with as few as three keystrokes; the system can be set up to turn itself on and transmit files at a preset time. The on-board RAM storage of the system will even allow exchange of messages when the computers at both ends are turned off.

A machine called the Ambiset follows a similar design, with features such as a speakerphone and call forwarding as well as a text processor and programmable calculator. With the Ambimail communications program, a user can send and receive text with one keystroke, set up unattended operations, and automatically forward phone calls.

Voice Mail

It is, in fact, no longer even necessary to have a computer with you to pick up your electronic mail. "Voice mail" is a store-and-forward equivalent of character-based electronic mail.

AT&T became the first major provider to offer such a service with its AT&T Mail system. Subscribers to that service can use a standard touch-tone telephone to call a central computer, and after identifying themselves through the use of a secret password, listen to a computer-synthesized voice as it reads them their mail. In this system, the computer is converting into speech-like sounds the text files put into storage from other computers.

Another form of voice mail uses the computer to answer the phone with synthesized speech and then record spoken answers for pickup by the intended recipient. The messages can be stored in an analog form on a cassette tape, or in a more sophisticated system the voice messages are stored in digital form on magnetic disk.

Voice mail offers great benefits in situations such as an order department's dealings with an outside sales force, or where there is regular voice communications from customers placing orders or requiring assistance.

Telex

Telex, an old technology that traces its roots all the way back to the early telegraph systems, is alive and well in the various corners of the globe. And, as an unexpected effect of the telecommunications revolution, the telex system has actually flourished in recent years as increasingly internationalized Western businesses have extended their reach into lesser-developed countries. Microcomputers have not yet landed on the desks of businesses and suppliers all around the world, and electronic mail ser-

The Electric Mailbox

vices have also not extended into all countries directly. But what has happened is that the major electronic mail systems, such as MCI Mail and EasyLink, have forged links to the existing telex system and its 1.6 million subscribers around the world.

In the US market, telex is, according to industry spokesmen, "mature," which means that it is no longer growing at a significant rate. But worldwide, the best guess is that telex will be growing in the low double-digit range for some time.

The flattening US market is still a fat one, with annual receipts of $150 million, and worldwide revenues estimated at $550 million. The largest market for telex is between Western Europe and the US. About half a dozen carriers, including MCI's Western Union International subsidiary (not to be confused with the Western Union Telex service, with which it competes), Western Union, and ITT, split the domestic market; overseas, most countries have government-owned monopolies called Postal Telegraph & Telephone companies running the show. Telex is also flourishing in lesser developed countries, in places like Latin America, the Far East, and the Middle East.

It is now easy to send a message from a desktop microcomputer to the telex system, and to receive incoming messages as if you were a part of that system. If your company will have regular dealings with overseas points, you should investigate the services offered by electronic mail suppliers in this area.

MCI Mail, which has been the most aggressive participant in the newly emerging electronic mail network industry, offers outgoing and incoming telex service from around the world and across the United States. To use the service, an MCI registrant merely enters a telex code as the address in the standard letter format for that service. Telex subscribers worldwide, whether or not they are registered for MCI Mail, can communicate with you by directing their messages to MCI's incoming telex number. The dispatches are placed in your electronic mailbox.

Western Union's EasyLink service is close to MCI Mail in range of services, including electronic mail, telex, telegrams, cablegrams, and a business, news, and sports database online. Western Union operates one of the largest telex systems in the world (its circuits are used by most of its competitors at one point or another in communication) and the company provides its subscribers with the phone book-like Western Union Telex Directory.

Master of Your Own Communications

ITT's Worldcom Telex includes access to real time "interactive telex." In effect, this allows you to use ITT's facilities for standard online communications. Ordinary store and forward telex is available, as is an electronic mailbox for incoming messages. RCA Global Communications offers the same range of services.

FUTURE TECH

Broadcasting Data

One of the true advantages of electronic mail is the fact that it is essentially an "asynchronous" or one-way communications system. The sender dials up and transmits his message at his convenience; the receiver signs on and accepts her message when she can. For the most part, however, current email systems are tied to the ubiquitous telephone. That may be changing in the near future, as new options in communications links avail themselves to the email user.

There is now a whole family of services that can deliver massive amounts of information, including electronic mail messages, over an electronic circuit. The information streams in through a wire, over a broadcast signal, or via one of several hybrid transmission media.

There are four developing technologies to keep an eye on here. Each adapts an existing high-speed broadcasting or narrowcasting channel: television signals, FM radio signals, cable television wires, and satellite transponders. In all but the last system, the data applications "piggyback" onto existing uses and equipment. All of the new forms of data transmission offer the opportunity for the quick, relatively inexpensive movement of tremendous amounts of information.

Television-based systems use the VBI, or vertical blanking interval, which is an unused gap in the video broadcast signal. The FM mode piggybacks data onto existing broadcast signals using a "subcarrier" or SCA signal. Cable television can have a secondary modulation of its signal for data, or specific channels can be set aside for high-speed, high-capacity transmission of data.

DBS, Direct Broadcast by Satellite, is made possible by the development of low-cost, small "dishes" that can link to communications devices hovering thousands of miles above us.

Let's take a brief look at each of these up-and-coming systems.

FM Subcarriers

We have all suffered through exposure to FM-based subcarrier signals for many years without realizing it. In most cases, the "elevator music" we hear in department stores, dentists' offices, and even from our phones while waiting "on hold," has been transmitted by Muzak or other companies using SCA piggybacked on an FM signal.

An FM transmitter encodes its information by modulating the frequency of its signal, the number of waves in a particular time period. (An AM station modulates the amplitude or height of its waves to send information.) A subcarrier signal typically sits on top of the audio part of the broadcast signal, and can use any type of modulation for its signal, including FM, AM, single sideband, and other methods. The user requires a special receiver to strip out the SCA signal and demodulate it into digital form if necessary; you can think of this device as an "FM modem."

Television VBI Signals

VBI systems recapture a great unused wasteland and put it to use to carry data. A television signal is a complex stream of information, with data on color, brightness, audio, and image. It is not a continuously flowing image; what the viewer sees is a series of still pictures that change slightly from frame to frame. The brain is fooled into thinking it is seeing continuous action.

Each "frame" of an American television picture (foreign standards differ slightly) is made up of 525 horizontal lines of illuminated dots against a dark background. The television's electron gun traces an image on the screen from the top left to the bottom right. When it reaches the bottom of the screen, it needs to zip back up to the top to begin again without disturbing the image already displayed. In order to protect the image on the screen, the equivalent of 21 lines of image are transmitted with the gun shut off. It is this moment of blackness that is called the vertical blanking interval.

The availability of this space is the result of advances in the technology of consumer television sets. New transistor-based televisions no longer need all 21 lines. Year by year, the Federal Communications Commission has been releasing more and more of the VBI lines for other uses, including the monitoring of broadcasting transmitters, advanced color adjustment signals utilized by some consumer television sets, for videotex screens, and for transmission of data.

Master of Your Own Communications

Merrill Lynch and IBM have joined in a new venture called Imnet to broadcast financial information over the nationwide Public Broadcasting System network. Data is sent from Merrill Lynch to a satellite, where it is retransmitted to the TV stations for inclusion in their local broadcast signal.

Direct Broadcast by Satellite

Direct broadcast by satellite may be the most dramatic means of transmitting information. Today's network of government and private communications satellites may each have dozens of transponders and be capable of handling hundreds of simultaneous telephone or television transmissions, and the current marketplace possesses overcapacity. Most of the United States is now inside the "footprint" of communications satellites. Some experts say that small satellite dishes two to three meters in diameter will soon cost less than $500, including reception electronics, making them comparable to telephone modems, FM SCA receivers, and other land-based systems.

How might such systems be used as part of a future electronic mail service? One answer was given in the plans for a new network that is a joint venture of the ABC Corporation and Epson America. This service will be what is called a "point to multipoint" system, allowing, for example, a home office to simultaneously send out messages such as price lists and inventory information to thousands of modems around the country.

Wireless Communication

Despite all of the advances in desktop computers and portables, there is still a binding chain for almost all users, the connection to the telephone circuit for telecommunications. The travelling salesman still has to stop at a motel or branch office to find a telephone line to hook his portable computer into; the roving reporter still requires a phone booth to transmit her stories, and the inventory or stockroom clerk still must return to a desktop to report on product. Even this, though, is in the process of change. The solution may come through the development and use of wireless modems that link a computer to the standard telephone system, or to a special-purpose network.

The ESTeem wireless modem is one early effort. This device can link as many as 255 computers or peripherals inside a factory or office or at a remote worksite. Transmitting at speeds up to 2400 baud, its signal can extend as far as one mile in a typical office or factory setting. Adding a special antenna can boost transmission distance to as much as 30 miles.

The Electric Mailbox

The wireless modem, though, is still limited by its reliance on a dedicated radio transmitter and receiver system; you cannot expect to travel the country and find base stations wherever you stop. The next step, already well along, is the forging of links from a portable computer to the cellular telephone network in use in most major US metropolitan areas. The cellular phone is the latest evolution of the former "mobile telephone" service for automobiles. The previous system relied on medium-power radio transceivers in mobile units that broadcasted to and received from a large base station antenna in metropolitan areas. The signal was often poor, and calls would fade in and out due to atmospheric conditions. Driving through the concrete canyons of a major city while talking on a mobile phone subjected both parties of the conversation to a chorus of beeps, buzzes, and hisses. And further, each metropolitan area could handle only about 1500 average phone calls per hour.

The new cellular phone systems use individual phone units that are less powerful, but considerably more sophisticated. Instead of seeking to communicate with a single large antenna, they work with many smaller antennas, each of which serves a small "cell" of space in a metropolitan area. The system is tied together by a computer system that switches signals from one cell to another as the mobile unit moves. The systems are capable of handling an average of 50,000 ordinary calls per hour.

Communications from portable computers to the cellular telephone system is accomplished with the assistance of a piece of hardware called a "bridge" that is actually a special-purpose modem designed to work with a cellular transceiver. At least one manufacturer has now brought to market a briefcase unit that includes a cellular telephone transceiver, a bridge modem, and a laptop portable computer, creating a completely mobile, battery-powered office. And, for those time when you must work in an aerial environment, many major airline carriers have begun to install telephone systems on board their aircraft. These systems are also capable of carrying data to and from portable computers for the executive who needs to be in constant touch.

From Here to Ubiquity

One thing is certain: electronic mail in one form or another is hear to stay. As more and more people discover its advantanges, the industry will expand to suit their communications needs. Though today over 100 million messages are sent annually over the public email networks, the electronic mail industry is in its infancy. Some industry experts predict that by 1990, this message volume will increase to over 800 million messages a year.

Master of Your Own Communications

If you who have yet to experience the convenience and efficiency of electronic mail, you have selected the right book. We will show you how easy it is to get started. But first a word of caution: Communicating by electronic mail may be habit-forming, but beneficial to the health of your business. Once you discover this new world of message exchange, you may wonder how you ever got along without it.

However, remember that this electronic mail business involves a changing market and an evolving technology. If you think that email is a godsend now, to use Al Jolson's prognostic line from his first talking picture, "you ain't seen nothin' yet."

Making the Connection

The term "electronic mail" has a new-tech, businesslike, almost mysterious ring to it. So much so, in fact, that many people who could benefit from email services avoid them. Even people who have a fair amount of experience with computer hardware and software balk at the thought of hooking up to transmit information to another computer. Actually, email is very little different from standard mail. It is just faster, and the sender and receiver of electronic messages have more control over the process.

But the process itself is not so new. We still start with the need to communicate. You need to tell Palmyra Products in Pittsburgh, PA to process your purchase order for peripherals, or you want to contact cousin Claude in Columbus to collaborate on the coming convention.

We still have to compose the message. With conventional mail you might write a note with pen and paper, dash off some quick lines with a typewriter, or use your word processor. Even telephone messages require a certain amount of composition before you make the call. You have to think about what you're going to say, who you're going to ask for when the phone is answered, and so on. To prepare an electronic mail message you will most likely use a word processor or text editor to write the message.

We still have to prepare the message for delivery. With conventional mail, this usually involves finding an envelope for your printed copy, writing the name and address of the addressee and your return address on the front, and sealing the whole thing up. The equivalent steps with an electronic mail service are all paperless and speedy. You'll have to prepare your message for acceptance by the service, enter an address electronically, and then order the system to send it.

Making the Connection

You have to select a delivery system for your message. With paper mail the most obvious choice is the US Postal Service, but as we've already discussed, there are alternatives, including courier services, overnight express systems, and facsimile transmission. With electronic mail the primary delivery system is the familiar telephone system and delivery is almost instantaneous.

And finally, there is the wait for a reply. With conventional mail or the telephone, the wait can be from several days to a week or more. With electronic mail the delay between sending your message and getting a reply may be only a matter of minutes.

So you see, there really is nothing so mysterious or formidable about electronic mail. You're already doing most of the things you'll have to do to use email services; the tools are just slightly different.

In this chapter we will talk about the hardware and software tools of electronic mail, how to select them and use them, and we'll help you understand some of the terms and processes associated with email to make the transition as painless as possible.

The Delivery System

There are many parts to any message delivery system. With conventional paper mail, many of these components are not obvious, but they function as part of the delivery process just the same. When you drop a letter in the corner mailbox, you set into motion a series of convoluted and complicated procedures that must be conducted in sequence before your letter arrives at its destination.

The letter must be picked up from the corner box and delivered to the local post office. There it is sorted and packaged with other letters going to the same general area, then sent to a central clearing site. Further sorting is done, it is put in another package, and sent off in a group to another office near its destination.

When it arrives at a distribution center, the bundle is broken apart, and individual letters are routed to local offices and then given to individual letter carriers who carry them to the mailbox of the recipient. Obviously the specific steps depend on the place of origin and the destination. It may be more or less complicated than we have outlined here. But you the sender don't really care about all these steps. All you want to know is that if you drop an appropriately addressed letter in the corner mailbox, will it arrive within a reasonable time at the mailbox of the addressee.

The Electric Mailbox

The same thing happens with electronic mail. The delivery system has a local beginning (like the corner mailbox), and a distant destination (like the mailbox of the addressee). In between, a myriad of complicated steps are undertaken to get the message from point A to point Z. Again, for the user, what happens in between doesn't really matter, as long as the system works and your message gets through.

With electronic mail, the sender has to handle a little more of the initial delivery process himself, though. You do, however, have the help of a computer and some sophisticated software.

The Corner Mailbox

You could consider your computer or terminal to be the corner mailbox of electronic mail. A terminal is a computer-like device with a video display, a keyboard, and a connection for a computer or other communicating device. Although a terminal may look a little like a computer, a terminal only reads information you type from the keyboard, displays those characters on the video screen, and sends data over a wire to a computer somewhere else. Early microcomputers used terminals and large computers still use terminals as the primary human interface. Although the word "terminal" is sometimes applied to intelligent devices that are really small, stand-alone computers, when we use "terminal" in this book, we are referring to keyboard/display devices. These are sometimes called "dumb" terminals because they are incapable of computing in the traditional sense. The terminal display screen is used to show us what the computer says, and we use the terminal keyboard to talk to the computer.

Most common microcomputers include the terminal parts, hardware and software that interpret information from a keyboard and display data on a screen, right inside the box. There may be an external keyboard and a TV-like display screen, but since the computer and the terminal are really one piece of equipment, there is no need for the separate computer connection required of a dumb terminal.

Interestingly enough, if you use a microcomputer as the basis of your corner mailbox for electronic mail, you'll also use communications software that makes it act more like a dumb terminal than a computer. And you may need a separate plug-in board or other attachment to provide the dumb terminal's connection to the outside world. This connection is usually called a "port" or an "interface" and it is used to hook up two computer devices. It can be used to tie a printer to a computer, for example, to link a dumb terminal to a computer, or, in the case of a microcomputer used for electronic mail, to connect two computers. This port is called a serial or

Making the Connection

RS-232 interface. To understand how this interface works, you first have to know a little about how a computer works.

A computer stores and transmits information in a very simplistic way. If you type the letter "A" on a computer or terminal keyboard, the character will be displayed on the screen, but to the computer what you typed was really 1000001. The computer keeps track of everything in terms of a series of ones and zeroes. One pattern of ones and zeroes stands for the letter "A," another for the letter "Z," and still another for the number one. With only seven digits (seven ones or zeroes) the computer can store and display all the upper case and lower letters of the alphabet, the numbers from zero to nine, and common punctuation marks.

In computer terms, each of these digits is called a "bit." Because we use seven digits (bits) to define the 128 most common computer symbols, we call this a seven-bit code.

In its simplest form, English is a 26-bit code. If you add lower case letters, you have a 52-bit code. With a 52-bit English code we can write letters, even whole books, but without punctuation, numbers, and other symbols.

The seven-bit code of ones and zeroes used by the computer to display and store the alphabet and other familiar symbols is called the American Standard Code for Information Interchange, or the ASCII code, or simply ASCII (pronounced AS'key). This is an agreed-upon standard used by most major hardware and software manufacturers. Now, don't worry; you don't have to learn to "speak ASCII." That's something your computer does for you. It is important to understand this aspect of the way computers handle text, however, to get a handle on some of the finer points of the computer-to-computer communication used in electronic mail.

When you write a message with a word processor, the computer stores the text as a series of data bits. When you want to send this computer-stored message to someone in the form of electronic mail, all of these bits have to get out of your computer and into the receiving computer. That is the job of the serial (RS-232) interface and its associated software.

RS-232 is simply a convenient name to put on the particular way two computers are connected for this type of serial communication. There are other serial connections used in computing, and they have different names. The most common one for electronic mail, however, is RS-232.

The serial designation with this interface refers to how the information is moved from computer to computer. In serial communications, bits of data are moved along a wire from computer #1 to computer #2 in a serial stream. Think of the data bits marching in single file, rather that rushing haphazardly like a mob.

If you think about that process very long, you'll see a problem. Once the stream of bits starts flowing over the wire, how does the receiving computer know where one letter or number stops, and the next one begins? The two systems have to agree on a couple of things first.

For one thing, both systems have to agree on how many bits will be used to define each character. Although only 7 bits are required to send most common characters, computers actually use many more characters than can be defined by seven bits. An eighth bit is used to define these extra characters that are used for graphics characters (rocket ships, smiling faces, arrows, and such), or special text attributes in word processing.

To help the receiving computer sort out the stream of bits, start bits and stop bits are sent on either side of each transmitted character. The transmitting computer must send data in the way the receiving computer expects to see it or the bit sorting process will go astray, and nothing but garbage will get to the other side. This type of communication is called asynchronous communication because the sending computer sends data blindly, but following certain guidelines. The receiving computer is on its own to sort out the received data. In some communications environments, notably with direct communication to some mainframes, synchronous communication is used. This requires slightly different equipment that is readily available, but you probably won't need it for most electronic mail applications.

The Software

A hardware interface handles the physical connection between computers. The job of selecting which data to send and of presenting the information to the serial port for transmission is handled by software. This software is called communications or terminal emulation software. Actually it does both communications and terminal emulation chores.

On the communications side, this software helps you select data from a disk file or accepts characters from the keyboard and presents this information to the computer's serial port for transmission. In handling the movement of data in this way, the software also makes your computer act like a terminal: you type on the keyboard and information is displayed on the screen and sent out the computer's serial port. That's exactly what happens when you use a dumb terminal.

Making the Connection

This software may also offer some other convenient features. Many packages, for example, help you build and maintain an electronic telephone directory so dialing a number is as simple as selecting it from an on-screen directory. Communications software may also include a programming language, of sorts, that lets you automate much of the communications process.

Here's how that might work. Let's think first about a conventional telephone call. You have to look up the number you want to call or recall it from your memory. Then you have to dial the number, wait for it to ring and for someone to answer. When the telephone is answered you have to determine if the answering party is the one you want to talk with. If it is, you say what's on your mind and hang up. If the wrong person answered the phone, or you dialed the wrong number, then you have to somehow get in touch with the right person.

We do this so automatically that it hardly seems a difficult task, but if you could attach a device to your telephone that would take care of all these preliminary steps and notify you when the call is ready, it would be a welcomed addition. Indeed, many busy executives rely on a secretary to handle these duties and they answer the phone only after you are on the line and have had to wait for them.

Essentially the same process takes place when you dial another computer to exchange information or to post electronic mail. This sequence of events is called the "login" or "logon" or "sign-on" sequence.

You have to dial the telephone, wait for it to ring and for the other computer to answer. Then you have to type a certain sequence of characters to tell the answering computer who you are and what you want. With most electronic mail services you have a unique identification code that must be entered when you first call up. This may be your name, or it may be a series of numbers.

Next, most computer connections require some kind of password. This prevents unauthorized access to the system and its facilities. All of this takes time and, if you are a typical user of electronic communications, you may have several telephone numbers and logon sequences to remember.

Many communications software packages let you pre-program all the telephone numbers and logon sequences you use regularly so that to call another computer requires only the selection of an item from a menu. A number of software suppliers even include many of these logon programs already written for you when you buy their product.

Data capture and file transfer are two other desirable and common features supplied with communications software. By putting your software in "capture" mode, any information that is displayed on the screen is also written to disk, stored in memory, or listed on the printer. This is a convenient way to receive your mail and keep it for future reference. Without the capture feature, after you look at an electronic letter on the screen, it is gone. Of course you may redisplay it, but you can't print it out or keep it in your computer's data files for future reference.

File transfer capability is a somewhat more sophisticated method of exchanging data between computers. When you are dealing with straight text or ASCII information, you can display it on the screen and capture it to disk relatively easily. All computer data is not stored in this manner, however. Computer programs, for example, spreadsheet information, and enhanced word processing files, may include types of information than can not be transmitted in ASCII form. In this case, we use a binary transfer routine with error checking to ensure successful data exchange. Error checking means that the software at each end of the computer link is talking to the other during the data exchange. Special procedures are used to let the receiving computer know how much of what kind of information is being transmitted and the receiving computer tells the sending computer what it actually got. In the event of a discrepancy, the last section of data is sent over and over again until it is transmitted successfully.

Some software suppliers use their own error-correcting protocol. In that case you will need the same software running at both ends of the link for successful error-correcting transfer. This could be a disadvantage if you must communicate with a wide range of services or individuals. Other packages use public domain routines such as Xmodem (sometimes called the Christensen protocol) or Kermit. The Xmodem protocol is widely used on public and private data networks and it provides reasonable data integrity. When selecting communications software, the Xmodem protocol is one feature to look for.

A newer protocol that is gaining acceptance is X.PC. The number of programs supporting this protocol is growing as it becomes a widely used standard.

The Hardware

A computer with an RS-232 serial port and communications software is not enough by itself to get you into electronic mail. You need one more piece of equipment: a modem. Modem stands for "*mo*dulator-*dem*odulator." It is used to convert the electronic data inside your com-

Making the Connection

puter into a form that can be sent across telephone lines. It takes the digital bits inside your computer and converts (modulates) them into an analog warble for transmission on the phone line. At the receiving end, the modem demodulates the warble back into digital bits for the computer.

The modem sits between your computer and the telephone line. An RS-232 cable connects the serial interface on the computer to the modem. A modular telephone jack on the modem accepts a wire from the telephone line. When your communications software sends data out the RS-232 port, the information is intercepted by the modem, changed into a form compatible with telephone transmission, and sent along its way. At the other end, the process is reversed. The receiving modem accepts the data from the telephone line, changes it into a form acceptable to the computer, and sends it on.

Some modems are designed to fit inside the computer and attach directly to the system's data and address bus, so they are appropriately referred to as "internal" modems. In this case, you don't need a separate RS-232 port.

Modems can be directly connected to the telephone line with a modular plug just like the one connected to your telphone, or they can be acoustically coupled. Direct connection devices are by far the most popular, though they may cost more than acoustic units. They also offer more features and smaller size.

Acoustic couplers have two rubber cushions or cups designed to accept the telephone handset. Inside the cups are miniature speakers and microphones that send and receive the audio tones that modems generate for data transmission. Acoustic modems are usually less expensive than direct connect devices, though with new technology the gap is getting narrower. And they generally transfer data at a slower rate than direct connect modems. With an acoustical coupler you lift the handset from the telephone and dial the number in the usual way, then place the handset in the rubber cups. Once the connection is made, these devices work just the same as direct connect units. An acoustic modem is a good choice when you're traveling because you can not always gain access directly to the telephone line. Most hotels and motels, for example, have hard-wired telephones because of the fear of theft, and an acoustic device is normally the only type you can use from a pay telephone. An acoustic modem is not a good choice, though, if you use one of the new "space-age" phones with odd-shaped handsets. Anything other than the old standard desk phone variety won't fit into the rubber cups on the modem.

The Electric Mailbox

Modems are available in a variety of configurations, from very simple devices that do little more than the most basic data conversion, to high-priced, intelligent models that store telephone numbers and logon sequences and provide security protection for your computer. Some even "pick up" your electronic mail for you when you are away from the computer or are working on another program. There are modems that support only asynchronous communication, only synchronous links, or both. For electronic mail applications you need a modem designed for asynchronous communication.

The basic features to look for in a modem are the ability to answer as well as originate calls, auto-dial, and speed. Originate-only modems are rare today, but there are still a few around. You don't pay much extra for the answering capability, and you'll probably get a lot more features with it.

Intelligent, or auto-dial modems, can accept a telephone number from your terminal keyboard or computer communications package, then generate dialing tones or pulses to make the connection. If you use computer communications much at all, this is a worthy feature. Some users even take advantage of this auto-dial feature to dial voice calls by keeping frequently-dialed numbers on their computer-based telephone directory. After the software and modem work together to dial the number, you tell the computer to let go of the line, then you pick up the handset and talk.

Hayes Microcomputer Products was one of the early players in the microcomputer modem marketplace. They had one of the first popular modems and communications packages, and many software and hardware vendors that followed chose to adhere to the Hayes intelligent modem command set. This is a desirable feature for your modem and communications software. Many manufacturers include a higher level of functionality in newer modems, but they maintain compatibility with the Hayes commands to ensure successful operation with a broad range of communications software packages.

You'll sometimes hear the Hayes commands referred to as the "AT" commands because all commands to a Hayes compatible modem begin with "AT" for "attention." To dial a number, for example, you would send "AT" for attention, then "D" for dial, and either "T" or "P" for tone or pulse dialing. This is the lead-in dialing sequence or the dialing prefix and it is followed by the telephone number. For example, "ATDT18005551212" would tell the modem to use tone dialing to call 800 directory assistance.

If you are using a dumb terminal you would enter the characters just this way to dial a number with a Hayes compatible intelligent modem. With a

Making the Connection

microcomputer and communications software, you would probably select the number from a menu or telephone directory and the software would generate the proper dialing prefix for you.

Telephone line computer communications are conducted at relatively slow speeds, compared to the speed of most internal computer operations. This is because the telephone line was designed for voice communications, and it simply can't accept data as fast as the computer is capable of sending it without losing information.

Three communications rates are common: 300 bits per second (bps), 1200 bps, and 2400 bps. The speed of data transmission is often referred to as the "baud rate," so when we talk about 300 baud, we mean a speed of 300 bps. Remember that it takes about 10 bits for each character you send (7 or 8 data bits, 1 start bit, 1 or 2 stop bits.), so the more bits you can transfer in a second the faster the data transfer can be conducted. You can get a rough idea of speed by dividing the bps rate by 10. At 300 bps you can transfer approximately 30 characters per second; at 2400 bps the rate climbs to 240 characters per second. To transmit a text file that contains 5000 characters would take about 2 minutes and 45 seconds at 300 bps, but only about 20 seconds at 2400 bps.

You pay for the extra speed in the cost of your modem, but for serious communications applications you'll probably want a modem capable of at least 1200 bps. Anything slower takes too much time for transfers and causes frustration. Remember too, that you can use the higher speed equipment at lower speeds. Most 1200 bps modems can automatically switch to 300 bps or lower after the connection is made.

You frequently will see among modem specifications the designation "Bell 212A compatible," or "Bell 103 compatible." These terms refer to the speed of the modem and say something about the precise communications protocol used. This term evolved from the names of two early Bell modem series. The "103" modem uses 0-300 bps asynchronous communications. The "212A" modem can use 0-300 bps or 1200 bps and, if it strictly adheres to the standard, also supports synchronous communications. Most microcomputer modems that claim to be 212A compatible do not include the synchronous capabilities.

You may also see modems that conform to the CCITT V.24 standard. This is an international 2400 bps protocol, and should also include synchronous capabilities.

The Electric Mailbox

Processing Words

Though most email services offer some sort of online text editor, a "full screen" editor generally makes life easier when you are composing messages. It can save considerable time and money if you will prepare your messages offline, then "upload" them to the email service. With this in mind, let's take a moment to consider word processors as they relate to sending email.

Computers at every level have been used to manipulate text, and that is still one of the most popular applications. In the early days, text editors were used by programmers to enter program code. These editors were primitive by today's standards, but they offered the most basic features of text handling that are still necessary. With text editors you can type information once, store it on a disk or diskette, change it, move it, and, in the case of electronic mail, transmit it to other computers anywhere.

Word processor selection is a highly subjective task. Each software package has its own personality. You should study that personality, and work with the package as much as possible before making a purchase. Fortunately, the prices on applications software are falling rapidly, so initial investment is not as great a concern as it once was, but you should consider how much time will be required for you to become proficient with your word processing software. If you spend hours learning to use one package, only to discover you have made the wrong choice, you have wasted time in addition to money.

To select a text management tool, you should first decide what level of performance you need. Very inexpensive (even free) text editors are available that will do everything you need if you mostly write letters, memos, or an occasional short report. They can be very easy to learn and use, and they generally produce text ready to send over an electronic mail link.

At this level you should consider such products as PC Write ($10.00 from Quicksoft, 219 First N. #224, Seattle, WA 98109), or Edit ($6.00 from PC-SIG, 1030 East Duane Avenue, Sunnyvale, CA 94086). Such "freeware" offerings provide more functionality than you would imagine, and they are easily applied in an electronic mail environment.

The word processor you already have will probably work fine to compose email messages, or you can investigate any of the dozens of popular, full-featured word processing packages available at the local computer store or by mail order.

Making the Connection

Assuming you find a package that does what you want, costs about what you want to pay, and is reasonably easy to learn and use, there is one other factor you must consider. For maximum flexibility in using electronic mail, your word processor should be capable of producing standard text or ASCII (American Standard Code for Information Interchange) files.

Many word processors, especially those with extended features like proportional spacing, variable margins, underlining, and special printer support, use characters that are outside the standard text character set. Some electronic mail and other online services can't handle these special characters.

Be sure your word processor has the ability to "print to disk" in ASCII text format. Instead of printing on paper with all the formatting codes needed to make your printer operate, you will need to select your program's option to print generic text to a disk file. You may need to tell the program you are using a "draft" printer for this purpose.

Remember, too, that some email services prefer to receive only a carriage return, and others prefer a carriage return and line feed. Many communications software packages can add or strip either a carriage return or line feed as required, but you should know which of these characters is used by your word processor when creating the output file. Also, be sure to specify a right margin that is less than the maximum line length allowed by the email service you are using. Normally, you will also want to specify left, top and bottom margins of zero.

Since many email services interpret certain characters beginning a new line as a command marker, be sure these characters do not begin a line in your output file. For example, The Source interprets lines beginning with a period to be command lines. Similarly, Quik-Comm uses the asterisk and EasyPlex uses the slash to indicate command lines. If you must begin a text line with these characters, you might try preceding them by a space. Since every word processing program is different, you will need to consult the user manual for your word processor for instructions on printing text to disk.

Going Shopping

Now that you have some idea of the features to look for in communications software and hardware, here is a "shopper's guide" to some of the many products available today. This list is by no means exhaustive, but this representative sampling should get you started in your search. The focus is on products designed to work with the most popular personal

The Electric Mailbox

computers, including the IBM PC and other MS-DOS machines, the Apple II and Macintosh, and CP/M machines.

COMMUNICATIONS SOFTWARE

Around, Teal Communications International, Inc., P.O. Box 8062, Blaine, WA 98230. Teal Communications, Inc., 126 East 15th Street, North Vancouver, BC V7L2P9, 604-984-4411.

Teal's Around asynchronous package operates on systems under PC/MS-DOS, version 2.0 or later. Around provides ample menu prompts and a text editor. It supports data capture, automated logon sequences, and a command language that includes condition statements. The Xmodem protocol is used for file transfers.

Pre-configured command files for several popular online services are supplied and these programs can be accessed with simple ALT key combinations. The Around command language can be used to create custom electronic mail systems among PCs running the program. Online help is available.

Around provides a set of sophisticated features that can generally be accessed easily by even inexperienced users. The lengthy menu sequences, which can not be bypassed, may become tedious for experienced operators, however.

ACSP, IBM Asynchronous Communications Support Program, IBM Corporation, 1133 Westchester Avenue, White Plains, NY 10604.

This IBM offering operates with PC-DOS computers and provides standard asynchronous communications features as well as pre-configured support for the VM/370 and TSO IBM mainframe environment. This is a no-frills program without the "bells and whistles" many PC users expect. Menus are Spartan, it uses no special display attributes, and it offers a relatively narrow range of features. On the other hand, ACSP is easy to use, it is functional, it is low cost, and memory requirements are minimal (64K bytes).

Given the range of full-featured communications software available, this product will appeal primarily to users who must exchange data and communicate in the IBM VM/370 and TSO environment. The pre-configured mainframe environment support makes ACSP particularly attractive to purchasers who must support inexperienced users on a mainframe link.

Making the Connection

ASCII Pro, United Software Industries, Inc., 1880 Century Park East, Suite 300, Los Angeles, CA 90067, 213-879-1553.

ASCII Pro is a versatile communications package with versions available for PC/MS-DOS, Apple Pro-DOS 3.3, and CP/M-80/86 systems. ASCII Pro includes a built-in text editor that is more flexible and easier to use than the MS-DOS EDLIN program, but it is similar in operation to EDLIN.

The communication speeds supported by ASCII Pro are somewhat broader than competitive offerings, from 50 bps to 38.4K bps. The higher rates (above 2400 bps) can not be used with most asynchronous modems, of course. Some large system modems can support these high transfer rates, and they can be useful for direct-connect PC-to-PC transfers. Many personal computers, however, are not capable of communications rates above 9600 bps.

The Xmodem, Kermit, and TopView communications protocols are supported for error-corrected data transfer. Popular terminals can be emulated, and an emulations parameters menu helps you change emulation features as necessary.

No telephone directory, as such, is available, although separate command files can be maintained for each number you dial. These files can store keyboard-generated macros. An unattended mode supports remote access to an ASCII Pro computer for remote file transfers, including the receipt of forwarded telephone and TWX mail.

BLAST, Communications Research Group, 8939 Jefferson Highway, Baton Rouge, LA 70809, 504-923-0888.

BLAST (BLocked ASynchronous Transmission) is a menu-driven, general-purpose communications program available for nearly 100 computers including IBM, Apple, CP/M-based systems, UNIX systems, and most minicomputers and mainframes. BLAST comes preconfigured so that it is compatible with your system.

BLAST supports error-free binary and text file transfers between computers running the BLAST software. This is probably the most popular application for this package, and its support for many computers and operating systems enhances its popularity. The software makes excellent use of an error-free duplex channel and can send and receive files simultaneously. Communications Research claims you can send and receive files up to 50 percent faster in this simultaneous mode over sending each file separately. If the data exchange session is interrupted during transmission, the software can pick up where it left off, avoiding starting from the beginning.

A terminal mode with text file capture and upload capability also is supported. In this mode, however, error handling and operator feedback are not as good. A 10K RAM buffer captures data during terminal mode operation and the captured information is written to diskette only when the buffer fills. If the target drive is not ready or the media is full, this data can be lost.

COMMX-PAC, Hawkeye Grafix, 23914 Mobile, Canoga Park, CA 91307, 213-348-7909

COMMX-PAC (communications exchange utility package) is a versatile and easy to use telecommunication program. Although it is designed to take full advantage of a "smart" modem's capabilities, it can be used with equally good results with an acoustic modem. Utilities included with the package support bulletin board operation and remote computer access for such applications as order entry. COMMX-PAC is a menu-driven program with well designed and easy to execute operations. No online help is provided, but the menu selections are explicit and the user manual directions are clear. Examples cover communication with major services such as Newsnet, file transfers from COMMX-to-COMMX systems, transfers with other systems, error-corrected transfers with Modem7/Xmodem, and using EasyLink and Telemail.

COMMX supports auto-dial, auto-answer, and auto-logon. The auto-dial directory can contain up to 700 entries, although creation of the directory is not as easy as some of the other COMMX-PAC functions.

Sign-on scripts are created in a macro file. When this file name is included in the phone directory sequence, the logon macro is automatically executed when the number is dialed. You can selectively bypass COMMX-PAC menus with macros or special startup instructions to save time after you become familiar with the system.

A proprietary COMMX error correcting protocol is used to send and receive files. Xmodem/Modem7 protocol is supported to receive data only. Data compression can be used to reduce the amount of information that must be transferred.

In the terminal mode COMMX can capture data to a disk or printer, but the printer must operate faster than the data transfer rate. Emulation of a wide range of popular terminals also is supported.

Making the Connection

CrossTalk and CrossTalk-XVI, Microstuf, 1000 Holcomb Woods Parkway, Roswell, Georgia 30076, 404-998-3998.

CrossTalk is both a terminal emulator and a file transfer program. It supports communication and file transfer with host computers as well as error-free data exchange with other CrossTalk users. CrossTalk is available for 8-bit CP/M systems, and as CrossTalk-XVI, for 16-bit computers under PC/MS-DOS.

The program is supplied pre-installed for many machines, but for those where the user must perform the installation, instructions are clear and precise. CrossTalk can be used with most modems, but comes set up for the Hayes Smartmodem or compatible modem.

Online help is always available. When you type "HELP" (or simply "HE") a complete list of all CrossTalk commands is displayed. "HELP" followed by a specific command presents complete information on that command. If CrossTalk asks a question that you don't understand, enter a question mark (?) and you are told what to do.

CrossTalk is basically a command driven program that gives experienced users direct access to the program's features. A nearly full screen prompt or status display is shown when the program starts. During communications this screen is erased, but you can call it back with a single keystroke as needed. Program functions, communications parameters, and modem commands are entered on a single command line at the bottom of the screen.

CrossTalk supports sophisticated host interaction, including a "learn" mode that automatically determines and remembers host responses, emulation of popular terminals, and a programming language for automatic connect and logon. A proprietary error correcting protocol is used for CrossTalk-to-CrossTalk data transfer. The package also supports Xmodem transfers. Some first time users accustomed to menu-driven software find CrossTalk somewhat imposing. Experienced computer users, on the other hand, appreciate the software's direct command access, quick response, and programming capabilities.

Framework II, Ashton-Tate, 20101 Hamiltion Avenue, Torrance, CA 90502, 213-329-8000.

Framework II is an integrated software package for the IBM PC and other MS-DOS computer. Framework version 1.0 included a copy of the Mite telecommunications program (see review below), but now Ashton-Tate

The Electric Mailbox

has integrated the telecommunications module into the Framework pull-down menu system. This makes accessing this function much easier, requiring only two or three keystrokes. A menu of services, such as CompuServe and MCI Mail, allows you to select the one you wish by pressing a number or moving the cursor, and it is possible to customize this menu to display the particular services you will be using. Setting up new parameters is relatively easy since your choices are displayed on the menus and sub-menus, but the module is not without quirks. (For example, we couldn't get the program to make the modem dial using 8 data bits, but it seemed to work without any problem using 7 data bits.) Documentation is scattered throughout three different manuals, making it difficult to find answers to such mysteries in a hurry or at all.

However, once you get through the setup stage and the keystrokes become second nature, it is surprisingly easy to use. The pull-down menus offer a responsive user interface that make it unnecessary to memorize a lot of commands. Probably the best thing about using an integrated package such as Framework II for telecommunications is the ease with which you may move information from an application, such as a spreadsheet, database, or word processing document, and back again. It is not necessary to exit your word processor to logon to an email service, or to logoff your email service to check a figure in a spreadsheet. A menu option allows you to output any document you are working on as an ASCII text file for uploading.

Xmodem, Batch Xmodem, Clink, CrossTalk, CRC and Hayes Smartcom error-checking protocols may be used to send and receive files. The program will emulate several popular terminals, including the VT-100, Televideo 920, IBM 3101, and others, and your keyboard may be remapped to fit your application. Of course, Framework II is too expensive to use only as a terminal program, but if you need an all-in-one program that includes communications capabilities, this package is worth considering.

Mail-Com, Digisoft Computers, Inc., 1501 Third Avenue, New York, NY 10028, 212-734-3875.

Mail-Com is a telecommunications program developed specifically for use with MCI Mail, although you may use it to access other services as well. Designed to run on the IBM PC and compatibles, including the PC AT, it requires 192K of RAM. The program includes a text processor that allows you to edit your messages offline, including text files created with other word processors. You may then send them with a single keystroke. You do not need to waste time waiting for MCI Mail prompts; the program will

Making the Connection

automatically logon for you, upload your messages, download your incoming mail, and logoff. This is not only a time-saver, but it could be a money-saver as well for those located in areas where communications surcharges apply.

Mail-Com will memorize your keystrokes at logon and replay the same sequence automatically for future sessions. In addition to sending and receiving text files, the program will transfer binary files, including spreadsheets, graphics and programs, between one Mail-Com user and another using Xmodem or its own proprietary error-checking protocol.

One of the most interesting features is its ability to send and receive messages after hours when the computer is unattended. An electronic log keeps track of all unattended activity. Mail-Com may be ordered online and billed to your MCI Mail account, or it may be obtained from MCI Mail representatives.

Mastercom, The Software Store, 706 Chippewa Square, Marguette, MI 49855, 906-228-7622.

Mastercom is a low-cost, full-featured telecommunication program that comes pre-installed for PC-DOS machines. CP/M-80 users, however, must set parameter values before the program can be used with one of the 100 or so systems supported. The software is offered with a money-back guarantee.

Mastercom is a menu-driven product that does not include online help. Menus are self-explanatory, however, and the manual directions are clear. The program supports auto-dial, auto-answer, and auto-redial when using a modem that has these features. Its host mode permits remote operation via modem. Files can be sent using error-correcting protocols when the remote system is using the Mastercom program or when the Christensen Xmodem protocol is used by both systems. Transfers are also possible without error checking using the Xon/Xoff protocol, and files can be received using the Xmodem CheckSum, CRC, or Xon/Xoff protocols.

Micro Link II, Digital Marketing Corporation, 2362 Boulevard Circle, Walnut Creek, CA 94595, 415-947-1000.

Micro Link II is a smart terminal communication program that is pre-configured for most CP/M and PC/MS-DOS computers. Unlike most programs that use alpha characters or mnemonics to refer to commands, Micro Link II assigns almost 50 numbers to identify commands. Some

The Electric Mailbox

numbers require further information. For example, to set line width to 80 columns you would enter "22.80."

Each time Micro Link II is loaded, you are given the option of viewing the Set Up portion of the program. Options permit changing settings, writing phrases for auto-dialing and logging on to services (for Hayes or compatible modems), and writing telephone numbers (for built-in modems).

Micro Link II uses four menus to present user options: Main, Screen and Character, Copy Action, and File Sending. The Main Menu is the first menu displayed and contains the most frequently used commands.

The program supports file transfer by Xmodem and Modem7 which makes it compatible with a wide range of other software. It can prepare word processing files for sending by deleting control codes except for the carriage return, line feed, and tabs.

The user's guide is well written, generally easy to use, and includes examples explaining how to use Micro Link II. The guide has a section devoted to trouble shooting that provides possible causes and solutions for a variety of problems. A good index is included in addition to an ASCII chart and glossary.

Microphone, Software Venture Corporation, 2907 Claremont Avenue, Suite 220, Berkeley, CA 94705, 415-644-3232.

MicroPhone is a new communications package designed especially for the Apple Macintosh computer. Authored by Dennis Brothers, who wrote an early "shareware" terminal program for the Macintosh called Mactep, MicroPhone is a full-featured terminal program offering many features normally found only on programs for the IBM-PC and similar systems.

The program allows you to create communications files for any service you may need to access, and it offers a choice of 20 baud rates from 50 to 57,600 bps. To avoid opening different files each time you logon to another service, you may use one file for several services if they use the same communications protocol settings.

Like Microsoft Access (see below) and some of the other second-generation communications packages, MicroPhone offers the ability to create script files for automated sessions and customization, and it will memorize your keystrokes and mimic them for future sessions. For transmitting and downloading binary files, the program uses Xmodem

Making the Connection

protocol and will transfer data in 1K blocks. It will allow your Macintosh to emulate TTY, VT52 and VT100 terminals.

Microsoft Access, Microsoft Corporation, 10700 Northup Way, Bellevue, WA 98004, 206-828-8080.

Microsoft Access is available for IBM PC and compatible systems under PC/MS-DOS, version 2.0 and later. Access goes beyond some communications and file transfer programs by providing a full-featured script language that can be used to automate logon procedures or to construct a custom operator interface.

Microsoft Access includes a text editor, a macro function, and supports both Xmodem and X.PC protocols. It is supplied with custom menus for several popular online services, and the Microsoft Access Script Command (MASC) language supports sophisticated end-user application development. MASC programs can include conditional statements (IF...THEN...ELSE) and mathematical functions. The language also supports memory variables and string manipulations. The program can support up to eight simultaneous communications sessions under X.PC. A menu structure similar to Microsoft's Multiplan spreadsheet package is used. Emulation for such popular terminals as the DEC VT-100, Televideo 910, and IBM 3101 is supported.

Mite, Mycroft Labs, Inc., P.O. Box 6045, Tallahassee, Florida 32314, 904-385-1141.

Mite was first introduced for CP/M machines in 1982 and was one of the first menu-driven communications programs. It is now available for PC/MS-DOS machines and the Apple Macintosh as well as for CP/M systems. It is available pre-configured for over 100 systems. Commands are mnemonic and easy to use. Mite's online help is easily accessed. A question mark (?) entered at the "Enter Option" prompt provides general help for an entire menu, or specific help for a particular item.

Mite supports most auto-dial modems and permits auto redial. Although a telephone directory is not available, up to ten pre-stored macro strings for semi- or fully-automated logons can be defined and stored as a part of the parameter files. Instructions for constructing auto logon strings are provided, and an example given. An unattended system can be set up using the auto-answer feature. A single password or a password file can be established to protect the host system from unauthorized callers.

The Electric Mailbox

Files can be exchanged with error checking with other computer systems running MITE and with systems using SmartCom, Crosstalk, and Clink. Xon/Xoff data transfers also are supported.

The user's guide is generally well-written, though it is slanted heavily toward the CP/M user. Users of PC/MS-DOS machines may have a little trouble with some of the instructions.

LYNC, Norton-Lambert Corporation, P.O. Box 4085, Santa Barbara, CA 93103, 805-687-8896.

LYNC functions under a variety of operating systems including PC/MS-DOS, CP/M-80/86, MP/M-80/86, TurboDos, Apple DOS, Concurrent CP/M, and Z-DOS. LYNC requires only 18K bytes of RAM.

This program supports logon script files to automate the logon process to public networks or company mainframes. A proprietary LYNC error correcting protocol can be used for transferring files between two LYNC systems. The Xmodem protocol also is supported.

A peer-to-peer connection between two LYNC systems permits remote access to another PC. You can access LYNC menus, execute programs, display disk directories, and conduct many other DOS-level functions from the remote connection.

An online user's manual is available from anywhere within LYNC and can be accessed by specific topic. LYNC can automatically determine the major protocol settings at the time of connection, setting the number of data bits, baud rate, and start/stop bits without user intervention. The program can be configured for communications up to 9600 bps, but the actual line transfer speed can be 1200 bps maximum. This is a problem only when two computers are directly connected, or when a new, high-speed modem is in use.

Move-It, Woolf Software Systems, Inc., 6754 Eton Avenue, Canoga Park, CA 91303, 818-703-8112.

Move-It is a communications package for a wide range of personal computers under PC/MS-DOS and CP/M-80/86. Terminal and error correcting modes are supported, but Move-It uses a proprietary protocol and the software must be running on both systems for error-free transfers. Xmodem, Kermit, or other public domain protocols are not supported.

Making the Connection

Although a disk-based telephone directory can be maintained, the program sets communications parameters during software configuration. It is not as convenient as with some other systems to change such parameters as communications port and baud rate.

Although Move-It operates on both CP/M and PC/MS-DOS systems, the documentation is slanted toward the CP/M user. The product is basically a command-driven package, and menus are sparse. User feedback is Spartan, and novice users may be initially intimidated by this package.

Move-It supports data rates to 19.2K bps and uses Hayes-compatible intelligent modem commands.

P/C Privacy, MCTel, Inc., 3 Bala Plaza East, Suite 505, Bala Cynwyd, PA 19004, 215-668-0983.

The "PC" in the name stands not only for "personal computer," but also "personal and confidential." This program is not a terminal program, but rather a program that scrambles your confidential data files, such as top secret email messages, Lotus 1-2-3 spreadsheet data, and even dBase II files, making them unreadable unless deciphered with the same program by an authorized individual. Encrypted files are converted to ASCII text that can be sent over most email systems. Both the sender and the recipient will need to have a copy of this program, of course, for the scheme to work. Versions are available for PC/MS-DOS, Apple DOS, and CP/M-80 systems.

PC-Talk III, The Headlands Press, P.O. Box 862, Tiburon, CA 94920, 415-435-9775.

PC-Talk III is one member in a growing list of user-supported programs for PC/MS-DOS machines. The so-called "freeware" programs are supplied over dial-up bulletin boards or hand-to-hand among users. There is no formal charge for the program, though Headlands Press asks for a $35 contribution from satisfied users. A small user's manual can be printed from within the program.

PC-Talk III is Hayes command compatible, offers an easy-to-use dialing directory and menu structure, and supports both Xon/Xoff and Xmodem protocols.

The directory holds up to 60 names, telephone numbers, and communications parameters. The program is pre-configured for the Hayes Smartmodem standard, but other modems can be user-configured. PC-

The Electric Mailbox

Talk III's low cost and friendly user interface have made the package one of the most popular asynchronous communications packages among PC/MS-DOS system users.

Perfect Link, Thorn-EMI Computer Software, 1881 Langley Avenue, Irvine, CA 92714.

Perfect Link is one in a series of offerings originally from Perfect Software, Inc. Thorn-EMI purchased the rights to the series in 1984, and Perfect Software has continued to enhance the products. Versions are available for PC/MS-DOS, CP/M-80, and the Apple II series. Perfect Link uses a series of menus and windows to step you through the communications process. Menu-selected utilities help you configure the system for various communications settings, and these configurations can be saved to disk for later use.

The package emulates popular terminals, including the DEC VT-series, IBM's 3101, the Lear Seigler ADM-3A, and the Televideo 920. The Xmodem file transfer protocol is supported.

Perfect Link includes what the company calls "Wireless File Transfer," a local-mode utility that lets you read files from one diskette format to another. For the systems it supports, this facility can save a lot of hook-up and communication time.

No programming language is supported, but Perfect Link can capture a series of keyboard commands and store the resulting sequence into a macro file for later retrieval. Overall, the package is easy to use, offers good menus and a friendly user interface. A strong point is the similarity of the user interface among other Perfect software modules. If you know Perfect Writer, Perfect Filer, or Perfect Calc, then this program will be very familiar to you.

PFS Access, Software Publishing Corporation, 1901 Landings Drive, Mountain View, CA 94043, 415-962-8910.

PFS Access is a relatively low-cost asynchronous communications package for PC/MS-DOS and the Apple II series. PFS Access supports communications at 0-300 or 1200 bps and offers simple TTY terminal emulation. Menu-oriented data encryption based on the National Bureau of Standards Data Encryption Standard (DES) is provided. This can be a useful feature for electronic mail transfers when privacy of communications is important.

Making the Connection

An automatic logon script for virtually any service or computer can be constructed by "recording" keystrokes to a named disk file. Up to eight of these logon scripts can be menu-selected. No formal programming language is provided. No error-checking transfer protocol is supported.

Novice users will appreciate the program's ease of use and clear menu structure. Experienced computer communicators likely will be disappointed by the relatively few features offered.

MCI Mail offers a product appropriately called MCI Mail Access, which is essentially the same program as PFS Access, but it has been customized to make it easy to logon to MCI Mail and check for your messages.

POSTPLUS, MCTel, Inc., 3 Bala Plaza East, Suite 505, Bala Cynwyd, PA 19004, 215-668-0983.

POSTPLUS is an all-in-one telecommunications and word processing program for CP/M and TRS-DOS systems. POSTPLUS supports smart or manual modems and acoustic couplers and accommodates auto-dial and auto-logon.

One key dialing and logon is available for 25 information services. POSTPLUS can also be used as a telex station. POSTPLUS is menu-driven and uses easy to remember single character mnemonics. Online help is reached by typing "H" at any menu. This takes you to the "help" table of contents from which you select the page number for the help item you need. In addition, you can create custom help files that can be accessed from the help menu.

POSTPLUS supports the UART protocols (word length, parity, and stop bits) which provide management of all major communications protocols. Such error-correcting protocols as Xmodem are not supported.

Telecommunications sessions can be saved in a file that can be reviewed, edited or printed. A separate text processing feature offers most standard word processing functions to help in preparing text files for transmission with the program's communications features. A split-screen feature lets you work with two text files simultaneously.

POSTPLUS can receive telex messages if you are using an auto-answer modem. Command sequences, logon instructions, and often used strings of text can be created in the User Area, and once in the file, they can be called out by a few key strokes for insertion in the text or for logging on to a favorite service. You can create as many macros as the disk will hold.

The Electric Mailbox

Confidentiality of information is possible with a feature called "scramble" which encodes files for privacy. (Also see PC Privacy above.) The user's guide is well written and thoughtfully organized, but the index lacks sufficient detail for easy location of material.

Qmodem, The Forbin Project, Inc., 715 Walnut Street, Cedar Falls, Iowa 50613.

Qmodem is a user-supported "freeware" communications package for the IBM PC and compatible computer marketplace. It is a fast, easy to use, full-featured communications package that will be especially familiar to users of PC-Talk III. The user interface is very similar to this popular public domain software, but Qmodem takes advantage of more of the PC's screen attributes, uses windows heavily, operates faster, and generally appears to be a slicker package.

Qmodem is supplied with a very useful, well-designed user manual of some 50 pages. All of the program's features are explained in an easily accessed reference format. A programming script language is included that lets you design auto-logon command files linked to entries in your telephone directory. To dial a number and logon to a host you select a telephone number from the directory window and the software handles the rest of the task.

Qmodem uses a series of ALT key sequences for most features, which makes program access fast and easy. If you forget any of the commands you can open a command window with the HOME key. You can exit Qmodem to DOS and leave the communications program resident. Qmodem supports a variety of error checking protocols, including Xmodem.

REACHOUT, Applied Computer Techniques, 104 Knight Drive, San Rafael, CA 94901, 415-459-3212.

REACHOUT, a program for CP/M systems, is menu-driven and remarkably easy to learn and use. Help is available from any prompt. Error-checking protocols supported by REACHOUT include checksum comparison, CRC comparison, and echo-plex. Dialing is accomplished manually from the keyboard or automatically by using the telephone number directory. Sign-on scripts can be developed as part of the directory, but can be cancelled by keying CTRL I before selecting the telephone number. Auto-redial and auto-answer are also supported.

Making the Connection

Files can be received or sent without error checking to information utilities, telex, or a mainframe system that "talks ASCII." Xmodem and Modem7 protocols are also supported. Batch files are permitted only when the other system is using REACHOUT.

Smartcom-II, Hayes Microcomputer Products, Inc., 5923 Peachtree Industrial Boulevard, Norcross, GA 30092, 404-449-8791.

Smartcom-II operates under PC/MS-DOS version 1.0 and later and with CP/M-80/86 machines. The Smartcom series was first introduced in 1981 as a natural progression from Hayes' earlier terminal program for the Apple II. The software obviously supports the popular Hayes Smartmodem command structure and provides a simple-to-use operator interface. The program is supplied with pre-configured setup files and macro command structures for easy access to a number of popular online services, including CompuServe, Dow Jones News/Retrieval, Knowledge Index and The Source.

Error-free communications and file transfers are supported between Hayes systems and machines that support the Xmodem protocol. A split screen display continuously shows the main menu at the top of the screen and transient functions, including a telephone dialing directory, on the bottom. The main limitation of the program is that it will not work with modems that are not completely compatible with Hayes hardware.

UAP-Link, Unique Automation Products, 15401 Red Hill, Suite G, Tustin, CA 92680, 714-730-1012.

UAP-Link operates under PC/MS-DOS version 1.1 or later. Versions also are available for minicomputers and mainframes. Users in DEC, Data General, and IBM VM/CMS environments with UAP-Link operating on the host can perform many host-resident tasks from a remote PC.

The package provides standard ASCII terminal emulation, but error-correcting protocols such as Xmodem and Kermit are not supported. UAP-Link is command driven and except for a short display showing host-based functions, no menus are used. A help screen lists program commands and prompts.

Sophisticated command files can be generated for use by the program for such tasks as autodialing, logging onto a host, transferring files, and other tasks. The Hayes modem command structure is supported.

The Electric Mailbox

VN Relay and Relay Gold, VM Personal Computing, Inc., 60 East 42nd Street, New York, NY 10065, 212-686-1450.

VM Relay and Relay Gold are companion, high performance communications programs that support asynchronous access for PC/MS-DOS machines. These products can be liked to VM mainframe software to provide full-featured access and error-free data transfers. The Relay protocol operates at higher speed than some other data transfer products. In TTY mode, Xmodem with either checksum or CRC error correction is supported.

Relay Gold also includes a high-level programming language that can be used to construct custom user menus, logon sequences, or shells for remote software access. A script language is provided in both systems for automating logon sequences. TTY and VT-series terminal emulation are supported. Relay includes a text editor that can be used while files are being transferred. The editor can be used to create or modify Scripts command files. The Relay products are menu-driven, and experienced users can bypass some menu screens if desired. Extensive online help is supplied.

MODEMS

AJ 347, Connection Modem, and 1212-AD1, Anderson-Jacobson, Inc., 521 Charcot Avenue, San Jose, CA 95131, 408-435-8520.

Anderson-Jacobson offers modems for just about any application. The product line ranges from low priced acoustic modems to commercial units operating at 9600 bps or higher.

The AJ 347 is a 300 bps device that can be acoustic coupled or connected directly to the telephone line. This could be a good choice for traveling computer users who sometimes find themselves in need of communication but can't access the telephone line. The unit supports auto-answer, but you have to dial the number yourself.

The AJ Connection modem, on the other hand, supports 300/1200 bps operation, stores telephone numbers internally, and will auto-dial the number for you. With this modem you can alternate between voice and data modes while you maintain the connection.

The AJ 1212-AD1 adds synchronous communications support to these features.

Making the Connection

Asher, Asher Technologies, Inc., The Quadram Communications Group, 1009 Mansell Road, Roswell, GA 30076, 800-334-9339.

Asher is a voice/data modem with multi-function software that includes support for up to five PC-DOS windows. The Asher package includes its own asynchronous communications software, and you can not use the company's internal modem with any other communications package. It is not Hayes command set compatible.

Asher supports concurrent programs running in the separate DOS windows. On a PC/AT or other high speed machine, these simultaneous prorams work rather well. On a standard PC, however, the multiple applications run painfully slow.

Asher includes software for communications, file card data management, calendar functions, and simultaneous voice and data communications. Two telephone lines are required if you want to talk and send data at the same time. A telephone handset is supplied with Asher so you don't need a separate telephone set to use the voice functions. You can use the card file database to dial your numbers and create auto logon command files.

Although the Asher system will function with only 128K bytes of RAM, operation will be severely limited. To take full advantage of Asher features a full 640K bytes of RAM are recommended.

Courier 2400, US Robotics, 8100 McCormick Boulevard, Skokie, IL, 60076, 312-982-5001.

The Courier is another in the growing line of 2400 bps modems targeted for the PC-oriented consumer market. It supports 0-300, 1200, and 2400 bps communication, auto-dial, and auto-answer.

DC-2212, Radio Shack, 1800 One Tandy Center, Fort Worth, TX 76102, 817-390-3921.

This direct-connect, 1200 bps modem is one of the few so-called "212A compatible" modems that actually supports synchronous as well as asynchronous communications. The DC-2212 operates at 0-300 and 1200 bps, supports auto-dial and auto-answer, and includes menu-driven firmware. Radio Shack also offers 300 bps direct-connect and acoustic coupled modems.

The Electric Mailbox

ERA 2, Microcom, Inc., 1400A Providence Highway, Norwood, MA 02062, 617-762-9310.

The ERA 2 communications package consists of a Microcom 1200 or 2400 bps modem and a smart terminal program for Apple II or IBM PC computers. The software, which is not available separately, supports emulation of popular terminals.

Microcom is best known for development of the MNC file transfer protocol, used by a number of modems and private networks, especially in the minicomputer and mainframe environment. The Microcom microcomputer modems support this protocol for error-free communications and file transfer. Other standard modem feature are supported, including auto-dial, auto-answer, and reverse dial.

IBM 5841/PC Internal Modem, IBM Corporation, 101 Paragon Drive, Montvale, NJ 07645, 800-426-2468.

These offerings from IBM provide an external, serially connected modem, and a half-card PC plug in card for internal use. Both modems support 1200 bps communications, respond to the Hayes modem command set, and can be switched between voice and data with hardware or software switches. They can automatically switch from tone to pulse dialing if the first attempt at tone dialing doesn't work. The external modem also supports synchronous as well as asynchronous communications.

The 5841 external modem can store up to 20 password-protected telephone numbers for automatic dialing, and each one can be tied to an internal logon sequence for completely automated attachment to a host.

Maxwell 1200V/1200PC, Racal-Vadic, 1525 McCarthy Boulevard, Milpitas, CA 95035, 408-946-2227.

Racal-Vadic's Maxwell series is a set of internal (IBM PC bus) and external modems for asynchronous communications at 0-300 and 1200 bps. The modems are shipped with GEORGE, Racal-Vadic's own communications software which takes advantage of some extended modem features. The modems also are compatible with the majority of the Hayes modem command set.

These modems include ROM-based software that provides a menu-driven operator interface for number storage and dialing. The external modem could be used quite well with a dumb terminal. The firmware provides fairly complete feedback on call status; it displays messages such as

Making the Connection

"Dialing," and "Ringing." It even tells you "VOICE!" if a real person answers the phone, but neither mode has a built-in speaker. You can tell a lot about call progress with a speaker that even imaginative software can't tell you.

You'll find one perplexing problem with the 1200V external model of the Maxwell series. It doesn't have a power switch, so it stays on all the time it is plugged in, and you can't disable the auto-answer feature. If the modem is connected to the telephone line and plugged into a power source, it will answer your telephone on the first or second ring.

The modems include internal self-software, and Racal-Vadic, a major player in the computer communications field, provides an excellent technical support and online diagnostics service.

POPCOM X100, Prentice Corporation, 266 Caspian Drive, Sunnyvale, CA 94088, 408-734-9855.

This versatile external 300/1200 baud modem will work with most any computer having an RS-232 serial port. The unit is relatively attractive as far as modems go, and it never gets in the way. The X100 hangs from the wall outlet, so there is no external power adapter necessary, and the modem does not take up any desk space. It supports the Hayes command set, but it offers an extended set of commands and error messages as well. The X100 allows you to switch from data communications to voice on the same phone line and then return to data transmission.

Self-test diagnostics automatically set the internal switches to coincide with the signals from your serial cable. The unit plays a "tune" through the built-in speaker when you plug the modem in the wall and when you connect it to your computer to let you know the unit is working and that the connection has been properly achieved, allowing fool-proof setup. However, should you have any problem, Prentice provides friendly telephone support. Prentice also makes an internal version of the modem for the IBM PC.

Security Modem, Cermetek Microelectronics, 1308 Borregas, Sunnyvale, CA 94088, 408-752-5000.

Cermetek has been producing modem modules for OEM manufacturers for some time. Now they're applying some of their own technology in a modem for the retail market. The Cermetek modem is Hayes command compatible and includes an automatic call-back feature to screen calls to your computer or bulletin board. Internal memory registers store modem

The Electric Mailbox

configuration settings and a list of authorized telephone numbers and user IDs. When the modem is in the security mode, all calls are terminated as soon as the caller enters an ID and password. The modem then dials the pre-set telephone number to re-establish communications.

This modem supports both synchronous and asynchronous communications sessions. Battery-backed RAM stores up to 16 dial-out numbers and 25 numbers for call-back verification. The Security Modem supports two telephone lines so you can use a separate line for the security call-back.

The modem is priced competitively for a device with all these features, but if you only need asynchronous communications in a Hayes-command compatible modem, other units can be obtained for less money.

Security Modem, ITT, Data Equipment & Systems Division, 20 Mayfield Avenue, Edison, NJ 08837, 201-225-6121.

This external, 1200 bps modem offers asynchronous and synchronous communications and such "smart" features as automatic tone or pulse dialing, call progress monitoring and reporting, and security features.

You can store passwords and call-back telephone numbers for up to 25 users in the modem. Security features include 8-digit password protection, same-line or second-line call-back. Built-in telephone line diagnostics also are included.

SmartModem 1200/2400, Hayes Microcomputer Products, 5923 Peachtree Industrial Boulevard, Norcross, GA 30092, 404-441-1617.

The Hayes modems set the standard for microcomputer asynchronous communications. The "Hayes Command Set" or the "AT" command set is used by most popular modems and communications packages. The Hayes 1200 and 2400 modems are desk-top units that communicate with your computer over an RS-232 link; the 1200B is an internal circuit board for the IBM PC and compatible systems and is sold with the SmartCom-II communications software.

The 1200 and 2400 baud units have similar appearance, but the 2400 does away with the familiar modem switches usually located under the front panel. The 2400 is fully programmable, and all features including speaker volume are set with "AT" commands from the computer or terminal. The 2400 also stores telephone numbers so you can auto-dial from a dumb terminal or with a simple communications package, and it supports synchronous communications.

All units support auto-dial, auto-answer, and reverse auto-dial. A 300 bps model, the Micromodem IIe, is available for the Apple II series. This modem is supplied with SmartCom I software.

SmarTEAM, Morrison & Dempsey Communications, 19209 Parthenia Street, Suite D, Northridge, CA 91324. 818-993-0195.

The SmarTEAM is an imported clone that looks and acts like a Hayes SmartModem 1200 modem. The SmarTEAM is a relatively new entrant to the United States, but it has been marketed successfully in Europe and the Orient for some time. The suggested retail price of the SmarTEAM is approximately half that of the Hayes counterpart.

The SmarTEAM functions well with any communications software that is true to the Hayes command structure. The SmarTEAM user's manual is small, but complete enough. Installation consists of connecting the telephone and serial interface wires and plugging the wall transformer into an AC outlet.

Smart-Cat, Novation, 20409 Prairie Street, Chatsworth, CA 91311.

Novation offers a variety of intelligent modems. The Smart-Cat series is a mid-range collection that is Hayes command compatible, and it is available as an external serial model or a plug-in circuit card for the IBM PC and compatible systems. The units operate at 0-300 or 1200 bps, and they are supplied with Mycroft Lab's Mite asynchronous communications package.

The Smart-Cat appears to fully support the Hayes modem command set. Both the internal and external models support auto answer and originate. A series of switches sets external modem configuration. The internal modem is pre-set for most settings, but you can configure it for any of the supported serial communications ports on the IBM PC. Dual telephone jacks let you connect a telephone set and the telephone line simultaneously. Novation also supplies synchronous/asynchronous modems.

VA212, Racal-Vadic, 152 McCarthy Boulevard, Milpitas, CA 95035, 408-946-2227.

There aren't too many synchronous modems priced for the consumer market. The Racal-Vadic VA212 is one of them. The VA212 handles standard

The Electric Mailbox

asynchronous communications well, but it isn't as flexible or full of features as some of the async-only products offered by other companies. It is not compatible with the popular Hayes modem command set, for example.

But it is a good crossover product between the high-priced, dedicated synchronous product common to the MIS environment, and the low-priced asynchronous-only modems common to the microcomputer marketplace. The VA212A has an intelligent user interface that uses menus and prompts to help you program the modem and dial numbers. All you need is a dumb terminal or a very simple communications program to get autodial and other features. The COMM.BAS program supplied with PC/MS-DOS is sufficient for full functionality of the modem.

The VA212A has a front panel with LCD display that lets you program telephone numbers and set the modem configuration without attaching it to a terminal or computer. This feature is particularly attractive where a central staff selects and programs hardware. You can store up to 15 numbers internally and dial them from the front panel or with commands from the computer. Two numbers can be linked and the second number will dial automatically if the first number is busy or doesn't answer.

Volksmodem, Anchor Automation, 6913 Valjean Street. Van Nuys, CA 91406, 818-904-9792.

Anchor is known for low-priced and functional modems for just about anybody's computer. The Volksmodem is a small (7.5 x 3.5 inch), light-eight plastic box powered by a nine-volt battery. It operates at 300 bps only and does not support "smart" terminal programs, auto-dial, or answer mode. You plug your telephone line into one jack on the side of the box, and a telephone set into another jack. Two switches select full/half duplex and talk/data mode.

The Volksmodem is a good choice to support your portable, laptop or small computer system when you're traveling and need access to an on-line service or electronic mail. This is a "universal" product that works with a variety of computers. You purchase a separate interface for specific computer hardware. The higher-priced Volksmodem 12 also supports 1200 bps rates and auto-answer.

Watson, Natural Microsystems Corporation, 6 Mercer Road, Natick, MA 01760, 617-655-0700.

Watson is an extraordinary communications product. It is an internal PC-compatible modem and sophisticated software that supports

Making the Connection

asynchronous data and voice communications. PC-Talk III software provides the traditional computer link, and Natural Microsystem's own voice digitizing package gives you interactive voice communications. Filer software stores your telephone directory and other information on disk.

You can digitize your own messages and store them on disk for after use. Watson also records messages from callers. It uses around 3K bytes of disk space for each second of voice storage, but the quality is excellent. You can program Watson to answer your telephone in your own voice, ask questions of the caller, and provide custom answers by responding to telephone touch tones. Optional software lets Watson call selected numbers in your directory and deliver a message, ask questions, and record answers. See the next chapter for a further discussion of this product.

Zoom Modem PC 1200, Zoom Telephonics, Inc., 207 South Street, Boston, MA 02111, 617-423-1072.

The Zoom Modem PC 1200 is an internal 300/1200 baud modem for the IBM PC and compatibles. This new entry in the modem market is compatible with the Hayes command set, but it offers features not found on the Hayes units. Of particular interest to email users is the RAM buffer that acts as a "data answering machine," which allows the computer to receive and store information, such as electronic mail, while the user works on another application. Also, if a speech synthesizer is plugged into the unit's audio input port, a caller could receive spoken information from the PC by pressing a few buttons on a touch-tone telephone.

Other niceties include a clock/calendar, "demon dialing" (continuous redialing of busy numbers), and several security features. The Zoom Modem can be programmed to wait for a touch-tone access code before giving a carrier tone, thus foiling hackers seeking active modems. The unit supports four COM ports, and since it uses the 16450 UART (rather than the 8250 UART used in many modems), the manufacturer guarantees that it will work with the PC AT and even faster AT clones. A 2400-baud upgrade card is available. Zoom Telephonics plans to release an external version of the modem soon.

AT&T and Other Voices

Voice Mail

Voice Mail is an area of electronic mail that is in its infancy, but one that promises to grow rapidly as technology and techniques mature. This form of electronic communication uses a combination of standard telephone equipment, computers or terminals, modems, and telecommunications networks.

Until recently the growth of voice mail has been slow, mainly because large, expensive computer systems were required for it, and probably because many users perceived voice mail systems as difficult to use. Also, the business community is largely based on written communication. A neatly typed letter on classy stationery is still considered by many to be the definitive business communication. Now that letter costs and delivery times have both increased, more and more companies do their business with the telephone. Handwritten memos or computer database entries record the transactions, but the message transfer is with voice, over the telephone. The perception and the technology of voice mail are changing, and some researchers are predicting a nearly 200% growth rate for voice mail systems within the next few years.

The specific features of each service vary, but generally voice mail falls into two types. One type records digital signals from your computer or terminal just like regular electronic mail. You can then access your messages electronically with a terminal, or you can have the mail service computer turn the electronic signals into speech so you can listen to your messages over a regular telephone.

A second type digitizes spoken messages and stores them on computer media in a manner similar to regular electronic messages. To use this kind of service you don't need a terminal or computer. You just call up a special telephone number and press buttons on a touch-tone telephone to tell the answering computer where to send your spoken message. The email (perhaps it should be called "V-mail") computer asks you questions in plain spoken language and you respond with the telephone buttons. When you're ready to deliver your message, you speak into the telephone as if you were talking on a tape recorder. You retrieve your messages in much the same way as with the first kind of service.

Both kinds of systems also are available for installation in your own company, though until recently all of these systems were based on expensive minicomputers or mainframes. Now microcomputer versions are available for under $500.00, plus the cost of your computer.

Computer to Voice Mail

AT&T Mail is an electronic/voice mail service of the first type and it is available for a modest fee to anyone with a terminal or computer. This flexible service uses the communications facilities built into the UNIX operating system, but it works quite well with non-UNIX computers or terminals.

If you're using a UNIX-based computer you already know that a UNIX network can handle electronic messaging by displaying a message on the screen when you receive something in your electronic mailbox. Depending on the application you're running, you can either open another screen window and read your mail with minimum interruption, or your can end the job you're into at the moment for a quick look at your mailbox. Either way, UNIX handles the majority of the task for you and the mail function is virtually transparent.

AT&T Mail takes this concept a step further. Once you're set up as an AT&T Mail subscriber, you can send electronic mail to UNIX systems that aren't on your local network. To do it, you simply dial a local access number through your computer modem, answer a few questions online, and either type in your message or upload a text file from your computer. The message is captured by the AT&T Mail computer, then routed automatically to the recipient's machine. Assuming the person to whom the letter is addressed has the computer turned on and a modem attached to the telephone line, the electronic message from AT&T Mail is dropped automatically to the recipient's computer where it can be read whenever convenient just as if it were sent over the local UNIX network.

The Electric Mailbox

You can use AT&T Mail with other communications equipment, but it becomes more like MCI Mail or EasyLink in that you have to call a central number to find out if you have any mail waiting.

But the real strength of AT&T Mail over other services is its ability to read your message to you over a standard telephone. Let's say you're on the road attending a meeting or making sales calls. You don't have your portable terminal with you, but you still want to keep up with your electronic mail. You call the AT&T Mail system, push a few buttons on the telephone when the computer answers, and the system digitizes your electronic message and reads it to you over the telephone. With voice delivery you can take advantage of all the strengths of electronic messaging--instantaneous delivery, retrieval at your convenience, low cost--without having to carry a computer and modem with you.

To send messages with AT&T Mail you need some special software on your PC. AT&T versions are available for most UNIX-based machines, as well as PC/MS-DOS machines and the Apple Macintosh. As with most email services today, this one is user friendly and simple to use. An online directory helps you find the address (electronic ID) of the person you want to contact, and once you've specified the recipient for your mail, the process is automatic and transparent. It doesn't matter, for example, if AT&T Mail has to send the message across the country and into another computer network, of if the addressee is across the street or down the hall. Neither rain nor sleet nor dark of night keeps the e-message from its appointed rounds.

This may sound rather esoteric and complicated, but actually the process is about the same as with any other computer-to-computer communication. You need terminal software, which AT&T supplies for a fee when you sign up for the service, and you need a modem and whatever cables or attachment your particular modem requires. That's it. The service uses asynchronous communications protocols and standard telephone lines.

You can also use AT&T Mail to send printed messages through he US Postal Service. In this mode, the service transmits your mail to the node nearest the addressee, prints the message on a high-quality laser printer, and mails it at the nearest post office. This provides quick delivery of mail at a low price to people who are not subscribers to the service.

The cost of this service is remarkably low. You can send most messages (even fairly long ones) for less than a dollar. A minute of voice mail is about half that, and a paper message delivered via US Mail costs around $2.00. Extra services such as same day or overnight printed delivery, return receipt on electronic mail, telex delivery, and COD messages carry additional fees.

Voice to Voice Mail

One way to achieve voice-to-voice electronic mail delivery is with a PC add-on device such as the one being marketed by Natural Microsystems, cleverly named Watson. It is a relatively inexpensive plug-in board for the IBM-PC and compatible systems that includes a Hayes-compatible modem, PC-Talk III communications software, database software, and facilities for answering the phone in your own voice and recording the caller's message. At first this sounds like an expensive versions of the telephone answering machines we all love to hate. But Watson is actually much more than that.

With Watson you can set up a series of voice messages that callers access by selecting from a spoken menu or by answering the questions that you supply. You listen to Watson's questions or menu choices and respond by pushing the appropriate buttons on your touch-tone phone. Watson can be set up as a private voice mailbox by establishing separate boxes for each user. You use the telephone pushbuttons to tell Watson who your message is for, and to tell the system who you are so you can listen to any mail you have waiting. Nobody else can listen to your messages unless you give them your private access code.

Watson also is being used as a sales tool, and to provide customer support. With some additional software you can program Watson to call everyone on a specified mailing list, deliver a message, and record a response. You could have it call everyone on your executive committee, for example, to ask if a meeting next Tuesday morning is convenient. The person you call is instructed to push one button for yes, another for no, and other buttons for additional responses (wrong party answered the phone, person called needs to talk to you directly, or the party wants to leave you a voice message).

Successful application of a product like Watson requires a good deal of careful planning, but if you set it up right, there's hardly anything that'll substitute for it. By providing natural sounding voice response, and the ability to respond to questions, you can expect much higher usage and a more positive user response than the old fashioned tape answering machine.

Watson digitizes your voice and the voices of any callers and stores the messages on a standard PC diskette or hard disk. The quality of Watson's

The Electric Mailbox

speech is quite good. Indeed, you'll be hard pressed to distinguish Watson's digitized voice from the original. The major drawback to this system, however, is its large storage requirements. It takes about 3K bytes of storage for each second of telephone-quality speech you record. That means for any serious voice messaging system you'll need 10 to 20M bytes of disk space dedicated to the Watson application.

Several other companies are jumping on the voice mail bandwagon, and as technology improves and competition heats up, the prices of voice mail hardware systems will surely continue to decline, while the applications for such systems can only increase. Here are some companies that are currently supplying voice mail systems and services:

Systems

Centigram
1883 Rainwood Street
San Jose, CA 95131
408-291-8200

Digital Equipment
146 Main Street
Maynard, MA 01754
617-897-5111

Natural Microsystems
6 Mercer Road
Natick, MA 01760
617-655-0700

Rolm Corporation
4900 Old Ironsides Drive
Santa Clara, CA 95050
408-986-1000

VMX, Inc.
1251 Columbia Drive
Richardson, TX 75081
214-907-3000

Wang Laboratories, Inc.
One Industrial Avenue
Lowell, MA 01853
617-459-5000

Services

AT&T Mail
1 Speedwell Avenue
Morristown, NJ 07960
800-367-7225

GTE Telemessenger
One Stamford Forum
Stamford, CT 06904
203-965-2000

National Phone Services
345 North Canal
Chicago, IL 60606
312-559-1111

Voicemail International
19225 Stevens Creek Blvd.
Cupertino, CA 95014
408-725-7800

The Source

The Source
Source Telecomputing Corporation
1616 Anderson Road
McLean, VA 22102
703-734-7500

The Source is one of the oldest and largest database services available to the general public. Founded in 1979, Source Telecomputing Corporation was purchased by The Reader's Digest Association, and in April 1983, Control Data Corporation purchased part ownership in the organization. The service has over 60,000 subscribers. In addition to electronic mail, The Source offers public bulletin boards, news and weather, investor services, travel information, conferencing, user publishing, online shopping and more.

SourceMail is an easy-to-use and versatile system that allows you to send and receive messages electronically. For paper mail, The Source also allows you to send Mailgram messages.

HOW TO SUBSCRIBE

You may purchase a SourcePak sign-on kit at most computer stores. The kit retails for $49.95 and includes instructions for obtaining your ID number and password, a three-ring binder, a member's directory, and the "Getting Started" section of the Source Manual. The balance of the manual is mailed to new members after purchase to assure that the information is current.

You may also subscribe by contacting The Source directly. Subscription information is available by calling 800-336-3366 or 703-821-6666. The SourceMail address for subscription information is TCA068. There is a

$49.95 membership fee, which entitles you to receive an ID number, password and The Source Manual. All fees for individual accounts are charged to the subscriber's Visa, MasterCard or American Express card. Companies may be billed directly with approved credit.

RATES

Daytime rates apply Monday through Friday, 7 a.m. to 6 p.m. From the continental US, Hawaii, Canada, and foreign countries, connect charges are $21.60 per hour. (Foreign subscribers may be subject to additional communications network charges.) From Alaska, the charge is $28.80 per hour.

Evening rates apply Monday through Friday, 6 p.m. to 7 a.m. and all day on weekends and holidays. From the continental US and foreign countries, the rate is $8.40 per hour, $13.20 per hour in Hawaii, $20.40 per hour in Alaska, and $10.80 per hour in Canada.

The above rates apply for 300 baud service. For 1200 baud service, there is a surcharge of $4.20 per hour for daytime hours and $2.40 per hour for evening hours. For 2400 baud, there is a surcharge of $6.00 per hour for daytime hours and $3.60 for evening hours.

Rates are based on the user's local time. For subscribers who are *not* in Daylight Savings Time areas, from the last Sunday of April to the last Sunday of October, daytime rates are in effect from 8 a.m. to 7 p.m., and evening rates apply from 7 p.m. to 8 a.m. at the user's local time. Charges are calculated to the nearest minute. On occasion, portions of the system may be down from 4 a.m. to 6 a.m. Eastern Time for maintenance.

There are storage charges for maintaining your own files on The Source. You are charged for each record on a monthly basis. A record equals 2048 characters. For the first ten records, the charge is 50 cents per record per month. Depending on the number of records you store, the charge can go as low as five cents per record per month.

There is a minimum monthly charge of $10.00 for all subscribers. Members may opt to pay a $95.00 annual membership, thereby saving $25.00.

ACCESSING THE NETWORK

Subscribers may access The Source through Uninet or Telenet. Canadian subscribers may access these networks through Datapac. Refer to the appendices for phone numbers of these networks. Certain areas may access

The Source

The Source's own network, appropriately called Sourcenet. For both 300 and 1200 baud service, the numbers are: Chicago 312-856-2112; New York Metropolitan Area 212-661-2510; Pleasantville/White Plains, NY Area 914-238-9299. In the Washington, D.C. area, the number is 202-448-9191 for 300 baud and 448-9440 for 1200 baud.

Subscribers located outside areas with network access may connect to The Source through a special WATS number within the continental US. There is a 25 cent per minute surcharge for this service. The number is 800-368-3343, or 800-572-3517 in Virginia.

CONNECTION PROCEDURES

The exact procedure depends on the network you are using. When you get your ID number and password, you will be assigned a system number. This is a number from 10 to 20 that designates which mainframe computer system houses your account. Before you sign on, you will need to have your system number, ID number and password ready.

Telenet, Sourcenet and WATS users must use a system code. Here are the codes for each system:

System	Code	System	Code
10	C 30124	16	C 301156
11	C 30138	17	C 301159
12	C 30147	18	C 301162
13	C 30128	19	C 301158
14	C 30149	20	C 301408
15	C 30148		

Sourcenet and WATS. Dial the local number and wait for the connection. You will see the word "Sourcenet" followed by the "@" symbol. Enter the appropriate system code. You will see the message indicating you are "Connected to The Source." Go to the section on "Logon Procedures."

Telenet. For Telenet access, dial the local access number and wait for the connection. When connection is made, press Return twice. You will see the "TERMINAL=" prompt. Type "D1." (Other terminal types are available. See the appendix for Telenet.) The "@" symbol will be displayed. Enter the appropriate code for your system. When you see the message "Connected to The Source," proceed to logon.

The Electric Mailbox

Uninet. For Uninet access, dial the local access number and wait for connection. At the "X" prompt, press Return, type a period ("."), then press Return again. The Uninet pad and port numbers will be displayed, followed by the "SERVICE:" prompt. Uninet service codes run consecutively from S10 through S20, corresponding to The Source system numbers. Enter "S" followed by your system number with no spaces. For example, if your system is number S12, type "S12" at the prompt and press Return. You will see the message indicating that you are "Connected to The Source."

LOGON PROCEDURES

When you are connected to The Source, at the ">" prompt, type "ID" followed by a space and your ID number. Press Return. You will be prompted to type your password. When you have entered your password, press Return again. For example, if your account number is XYZ123 and your password is "CABBAGE," you would type the following:

>ID XYZ123 (Return)
Password: CABBAGE (Return)

For security reasons, your password is not displayed. If your account number is valid, you will see a welcome message and a copyright notice.

USING SOURCEMAIL

If you are new to The Source, you may prefer to use the menus they provide. These will give you a selection of services. But as you become more experienced with using The Source, you will find that it is faster and more convenient to simply enter word commands from the keyboard. So that you may become comfortable with the commands used for electronic mail, all lessons in this chapter will be executed in the Command Mode.

When you are logged onto The Source, enter Command Mode by typing "Quit" at any menu prompt. When you see the "->" prompt on the left side of your terminal screen, you are in the Command Mode. (Later in this chapter, you will learn how to bypass the menus entirely and go straight to the Mail program when you logon.)

Getting Help

Remember that help is available at most prompts on The Source. So, if in doubt about the commands you can use, try typing "HELP" and pressing Return. In most cases, a list of available commands will be printed on your screen. For information about a particular command or program, at the Command prompt, type "HELP" followed by the command. For example, "HELP MAIL" will give you information on SourceMail. For general information about The Source, type "INFO" at command level. You will receive a menu of options from which you may select the type of information you need. This online service is free.

Special Control Characters

Knowledge of the following control characters will help you manage your online time better.

> CTRL S   Stops the scrolling of text.
> CTRL Q   Resumes scrolling of text.
> CTRL P   Interrupts the display. You will be taken to
> (or Break)   Command level or to the main prompt.
> CTRL H   Backspaces.

The Source offers a feature called Chat that allows users to page other users for an online conference. When you are paged for Chat, a beep will sound and a message will appear on your screen telling you that you have been paged for a Chat. If you wish to chat with another user, consult The Source Manual for instructions. Suffice it to say here that the paging message can be annoying if you are trying to upload files to send via SourceMail. Thus, you may wish to turn off the Chat feature by typing "CHAT -OFF" at the Command prompt. This will prevent your session from being interrupted by Chat requests. You may reinstate the Chat feature by typing "CHAT -ON."

SENDING A LETTER

Sending SourceMail is very simple, and several convenient sending options are available to you. To send a letter, at the Command prompt, type "MAIL SEND" and press Return.

> ->MAIL SEND

At the "To:" prompt, type the ID number of the person to whom you are sending the letter, and press Return.

> To: BT1773

You will be prompted for a subject. Type a line of text up to 32 characters in length describing the nature of your letter.

Subject: TEA PARTY

Next, you will be prompted to enter the text of your letter. At this point you may type the text or upload a previously prepared text file. Be sure that each line ends with a carriage return and that the text on each line does not exceed 78 characters. Otherwise, portions of your message may be lost. Also, as you will see later, no line should begin with a period unless you are issuing a command.

->SEND MAIL

To: BT1773
Subject: TEA PARTY

Enter text:

Dear Mr. Revere,

You are cordially invited to a Tea Party to be held in honor of His Majesty King George III at Boston Harbor on the Evening of December 16th. All guests are requested to appear in native costume. I trust it will be a memorable evening for all.

Regards,

Samuel Adams

At this point, you may wish to use some of the sending options available to you. To do this, you may issue "dot commands." These are commands that begin with a period on a new line. Each command should be on a line by itself, preceded by a period and followed by a carriage return. For example, to issue the "Blind Carbon" command, you would type:

.BC XYZ123 (Return)

When you issue a dot command, you will be prompted to "Continue," which means you can enter more text or another command. Here is a summary of your sending options:

AR *Acknowledgement Requested.* Sends you an anknowledgement indicating the time and date the letter was received. If you follow the command by a space and an ID number, when sending a letter to more than one recipient,

The Source

you will receive an acknowledgement from that user. Otherwise, you will receive acknowledgements from each recipient.

BC   *Blind Copy.* Sends a blind copy of the letter. Follow the command by a space and an ID number. That user will receive a copy of the letter, but the original recipient will not see that the copy was sent. To verify that the entire text of your message was properly included in the letter, you may want to send a copy to yourself. To do this, simply use your own ID number after the command.

CC   *Courtesy Copy.* Sends a copy of the letter to one or more users. Follow the command by a space and one or more ID numbers. The original recipient will see that the copies were sent.

DI   *Display.* Prints the entire text of your letter on the screen.

DI SU   *Display Subject.* Displays the subject line of your letter.

DI TO   *Display To.* Displays the "To:" address line of your letter.

ED   *Edit.* Allows you to edit the text of your letter using the Source online editor. Consult the Source manual section on "Files and Features" for a list of editing commands. You must type "FILE" to end editing.

EX   *Express.* Places the letter at the top of the recipient's mailbox so it is displayed first when mail is read. This command is useful when you are sending more than one message to the same user and you wish to have one of the messages read before the others. To send an Express letter to selected users, follow the command by a space and an ID number.

LOAD   *Load file.* Loads a file into the text portion of a letter where the Load command appears. Follow this command by a space and the name of a file you have created on The Source. Refer to the Source Manual section on "Files and Features" for information on creating files.

NOSHOW   *Don't show list.* Suppresses the display of the list of recipients when the letter is read. Unless you need for

The Electric Mailbox

each recipient to see the list of ID numbers to which the letter was sent, you should use the Noshow command so that each recipient does not have to sit through a display of the list when they read your letter.

PASSWORD   *Password protect.* Requires the recipient to enter a password before the letter can be read. Follow the command with a space and a word that only you and the recipient know. This feature can be used for private messages when more than one user shares an ID number. Be sure the recipient knows the correct password because the message cannot be retrieved without it.

QUIET   *Quiet.* Suppresses verification of delivery. Normally, when you use the Send (.S) command, as each copy of the letter is delivered, a message is displayed verifying the delivery. If you are sending a letter to a long list of users, and the system is busy, this can be time-consuming. Unless you are in doubt about the validity of a certain ID number and need verification of delivery, you may want to use the Quiet command so that you can log off immediately after sending the letter and save connect charges.

Q   *Quit.* Cancels the letter and takes you to the mail prompt.

S   *Send.* Sends the letter. This must be the last command entered.

SU   *Subject.* Changes the subject line of the letter. Follow the command with a space and a new subject line up to 32 characters in length.

TO   *To.* Adds users to list of addresses. Follow this command by a space and one or more ID numbers separated by spaces.

Many of the sending options can be used at the "To:" prompt. When used in this way, do not enter the period (".") before the option.

To send the letter, type ".S" on a new line and press Return. The Send command should be the last command you type. This tells SourceMail to deliver your letter. No options or text can be entered after the Send command. Once a letter is sent you cannot "unsend" it; only the addressees can retrieve it. Unless you used the Quiet command, The Source will

The Source

display a message saying that the letter has been sent to each ID number in the address. For invalid ID numbers, the system will give you the option to delete or correct the ID number.

After the letter is sent, you will see the SourceMail prompt, which looks like this:

    <S>end, <R>ead, <SC>can, <D>isplay, or <Q>uit?

To send another letter, select "S." To read or scan, select "R" or "SC." The Display selection allows you to view any letter in your box. Selecting "Q" for Quit returns you to the Command prompt.

USING MAILING LISTS

SourceMail is a very economical communications medium if you need to send the same message to many different subscribers. At the "To:" prompt, or after the CC, BC, AR or TO commands, you can specify several account numbers as long as each account number is separated by a space and the line of numbers does not exceed 78 characters. For example:

    .TO XYZ123 ZYX321 YZX132 ZXY231
    .CC ABC321 ABC123
    .BC XXX321

However, if you regularly send messages to the same list of correspondents, you can create a mailing list (called a "distribution list" on The Source) and store it in your file storage area. You can create mailing lists online or you can create a text file with your word processor and upload it to The Source. Mailing lists avoid the need of typing or uploading account numbers each time you send a letter. The Source charges a fee for file storage. (See the section on Rates above.)

The Source provides two methods for preparing mailing lists: a "MAIL.REF" file and a text file. The most versatile method is to prepare a text file for each mailing list you want to use. These types of files are simple to prepare offline and can easily be updated. We will show you how to prepare a text file for a mailing list. For information on preparing a "MAIL.REF" file, consult The Source Manual.

To create a text file online, type "ENTER" followed by the name you want to give the file. You will be prompted to "Enter Text." Type each user's ID on a separate line. Normally, you will want to use the Noshow and Quiet

The Electric Mailbox

commands when sending a letter to a large number of recipients. Refer to the list of sending options above. End the list by pressing Return twice. Your list will look like this:

>ENTER SALES

Enter Text
XYZ123
ZYX321
ABC321
ABC123
(Return)
(Return)

To use this list, at the "To:" prompt, you would type "SALES" as the ID. A copy of your letter will be sent to all users in the list with valid ID numbers.

Mailing lists can be updated using The Source Editor. See the chapter in your Source manual titled "Files and Features" for details. To save online time, consider preparing your mailing list using your word processor and text editor. Be sure that there are no blank lines or formatting symbols in the file, there are no blank spaces on the left margin, and each account number is followed by a carriage return. When your list is complete, you can upload it to The Source following directions in the following section on uploading files.

UPLOADING FILES TO THE SOURCE

Uploading files to The Source is easy, but there are certain restrictions. Each line of text must not exceed 132 characters, but if you transfer files to be used in SourceMail, each line is limited to 78 characters. Only ASCII text files are supported.

The Source provides two procedures for transferring files: FILETRAN and RCV. Use RCV if your communications software does not support flow control.

Filetran

To transfer a file using Filetran, at the Command prompt type "FILETRAN." You will be asked to enter the number of lines you want to transfer before seeing an acknowledgement. Enter this number at the prompt and press Return. Next, you are prompted for a filename. Type the name you wish to give the file. At this point, take whatever actions your

The Source

software requires to initiate the transfer. When the transfer is complete, use your Break key or press CTRL P. You will be returned to command level.

RCV

To transfer a file using the RCV command, type "RCV" at the Command prompt. You will be asked for the name of the file to save the transferred file in. Type it, and then begin your transfer, taking whatever action is required by your software to send a file. You will see a question mark ("?") printed at the beginning of each line of transferred text. This will not show in your transferred file. When transfer is complete, type "$$DONE." at the last question mark.

It's a good idea to check each file transferred to make sure the format was not altered. You can do this by typing "CRTLST" followed by the filename at the Command prompt. To edit the file using the system editor, type "ED" followed by the filename.

If you want to include a file in a letter or other document you are preparing, use the ".Load" command, which is described above in the section on Sending Options. You may begin your letter online or simply load your file. To load a file, type ".LOAD" followed by the filename. You will be informed when the load is complete and prompted to continue. At this point, you may load another file, continue typing your letter online, or enter the Send command (.S). You will not see the loaded file on your screen. If you want to be sure the recipient received the letter in the format you intended, send a blind copy of it to yourself.

FILES AND FOLDERS

Every Source member is provided with two major filing areas called "UFDs" (for "User File Directories"): a main account UFD that is only available to you; and the Sharefiles UFD, where files can be stored and shared with other members. A number of commands are available allowing you to manipulate these files. Some of the commands are listed below. Be sure to consult the chapter "Files and Features" in your Source handbook for full details.

ED filename.	Edit a file.
DEL filename.	Delete a file from your UFD.
TY filename.	Display (type) a file.
CRTLST filename.	Display a file, pause every 24 lines.
F	Files. Display a directory your files.

The Electric Mailbox

READING MAIL

When you have logged onto The Source, at the Command prompt, you have three options that allow you to see if there is mail waiting in your box. First, to see if any mail is present, type "MAILCK" and press Return. This Mail Check command will display how many messages are in your box. For example, if you have two messages that have not been read, one that has been read, plus one express letter that has not been read, The Source will display the following message:

> 1 read, 2 unread, 1 unread express, 4 Total

This is a quick way to see if you need to read your mail. The second option is the Scan command, which gives more detail about the messages in your box. Each message will be numbered and include the heading, showing the ID of the sender of each letter, the date and time sent, and the subject. To scan your mail, type "MAIL SCAN" and press Return. When the heading for each letter has been displayed, you will be prompted to "Read or Delete by number." If you wish to read certain letters, you can type "R" followed by a space and the numbers of the letters you wish to read. Each number should be separated by a space. For example, to read letters 2 and 4, you would type "R 2 4" and press Return. To read all letters, you can type "R ALL" and press Return. To read messages up to and including a certain message number, you can type "R -" followed by the message number.

To delete letters, type "D" followed by the numbers of the letters you wish to delete. Typing "D ALL" will delete all messages in your box.

To actually read your mail, type "MAIL READ" at the Command prompt, and press Return. Letters will be displayed one at a time, and you will be prompted to dispose of the letter is some way before the next letter is displayed.

Note that letters are displayed in reverse to the order in which they were sent. In other words, the most recently sent letters are displayed first. Express letters are displayed first, then regular letters. If you wish to read letters in a particular order, you should use the Mail Scan command instead of the Mail Read command.

When 24 lines of data has been displayed a "--More--" prompt will appear to indicate that there is more data to read. To continue, simply press Return. (As you become more familiar with using SourceMail, you may want to enter commands at this prompt to speed up the process of reading

The Source

your mail. The options available at the Disposition prompt are also available at the More prompt. Refer to the Save command in the section on Read Options below for information on reading documents without the More prompt.)

Read Options

After each letter is displayed, you will see the "Disposition:" prompt. The Source is asking what you would like to do with the letter. Here are your command options:

AG *Again.* Allows you to read the entire letter again.

D *Delete.* Deletes the letter from your box and displays the next letter in your box.

F *File.* Files the message in a special file called MAIL.UFD. This removes the letter from your mailbox but allows you to read the letter at a later time. At the Command prompt, you can type "MAIL READ FILE" to read the letters you have filed.

FO *Forward.* Forwards a copy of the letter to another recipient. Follow the command by a space and an ID number. You will be prompted to "Enter text:" so that you may add comments to the beginning of the letter. For example, you may wish to type, "Bill: Thought this message might interested you. Regards, John." Use the Send command (.S) to send the message.

H *Help.* Lists commands available to you.

N *Next prompt.* Takes you to the Disposition prompt when used at the More prompt.

NE *Next letter.* Displays the next letter in your box without prompting you for Disposition.

P *Previous prompt.* Takes you to the previous prompt.

Q *Quit.* Exits SourceMail and takes you to the Command prompt.

RE *Reply.* Allows you to reply to the just-read message. You are prompted to enter text. You may use the sending options by typing the appropriate dot commands within your text. When you are finished, use the Send command (.S) to send your reply. Unless you

The Electric Mailbox

change the subject line with the .SU command, the subject will be displayed as "Reply to:" followed by the original subject.

SA   *Save.* Saves a letter in a file on The Source. Follow the command with a space and a filename. To read the file at a later time, at the Command prompt, you would type "TY" (for Type) followed by a space and the filename. You may wish to save only the text of the letter, in which case you would type "SAVE TEXT" followed by the filename. To add to an existing file, use the word "IN" before the filename. For example, to save the text of a letter to an existing file called "SALES," you would type "SAVE TEXT IN SALES". If you are receiving a long letter that contains a large amount of data that will be downloaded and reformatted for other use, such as a mailing list or table of information, you may want to use the Save command so that you can display your file without More prompts. This also saves having to press Return at each More prompt.

UN   *Unread.* Places the message in your "unread" stack.

Note that after entering the Save, File, Forward, Reply, or Help commands, you are returned to the Disposition prompt. This gives you the opportunity to enter another command or to press Return to display the next letter.

SHORTCUTS

When you get comfortable with SourceMail commands, you may want to combine or abbreviate commands to make sending and receiving mail even faster and easier. Here are some options you may want to use.

When sending a letter, you can specify many of the sending options in the Mail command at the Command prompt. In fact, you may want to put this command string as the first line of your word processing file and upload the file at the Command prompt. Here is an example of a command string that could be entered at the Command prompt:

MAIL EX AR XYZ123 ABC456 CC ZYX321 SU Sales Reports

This command string, followed by Return, would tell SourceMail that you want to send an Express letter, with Acknowledgement Requested from each recipient, to XYZ123 and ABC456, plus a Carbon Copy to ZYX321, using the subject heading of "Sales Reports." This bypasses the "To:" and "Subject:" prompts, so you are only prompted to enter text. (If you use the SU option in the command string, it must be the last item in the string.

76

The Source

Also, remember that your command string should not exceed 78 characters.) If you want to use other options or add addressees, you can use dot commands within your letter.

There are several options to allow you to selectively read or scan your mail without having to respond to the normal prompts. Also, you can abbreviate the Read and Scan commands as "R" and "SC" respectively. The following command strings can be entered at the Command prompt:

MAIL R UN	Read unread messages.
MAIL R EX	Read express messages.
MAIL R SU	Read messages with a specific subject line. Follow the command with a subject line.
MAIL R DA	Read messages sent on a specific date. Follow command with a date in this format: 31 DEC 1986
MAIL R FROM	Read mail from a specific sender. Follow command with a user ID number.

The above options may also be used with the Scan command. Just subtitute "SC" for the "R" command in the command strings. If you include the File command after the "R" or "SC" command in the string, it indicates that the command applies only to the letters you have placed in your MAIL.UFD file using the File option when you read your mail.

You can check your mail automatically at logon by creating a "C_ID" file. This is a file containing Source commands that you wish to have executed as soon as you are connected. To create a file, at the Command prompt, type "Enter C_ID" and press Return. Type each command to be executed on a separate line.

For example, if you wanted to turn off the Chat function, so as not to be interrupted while looking at your mail, and to scan your mailbox automatically, you would type the following commands:

    CHAT -OFF
    MAIL SCAN

To close the file, press Return twice. To verify that the file has been saved, type "F" at the Command prompt to see a directory of your files. The next time you sign on to The Source, these commands will be executed automatically. To cancel the commands and get to the Command prompt, press Break, or type "Quit" at the first prompt.

The Electric Mailbox

When you logon to The Source, you can type your password on the same line as your ID. Just type a space after your ID number and then type your password. For example, at the ">" prompt, if your account number is XYZ123 and your password is "CABBAGE," you would type:

>ID XYZ123 CABBAGE

Be aware that your password will be displayed if you type it on the same line as your ID, so, for security reasons, it is not advisable to do this if there are people looking over your shoulder.

SENDING MAILGRAMS

As a Source subscriber, you may send paper mail by using the Western Union Mailgram service. The Source charges $5.15 for a single Mailgram of 100 words or less. Rates increase based on the number of words in the message. When sending the same message to more than 25 addresses, the cost drops to $3.50. For a complete current list of Mailgram charges, select the Rates option from the Mailgram menu.

To access the Mailgram menu, at the Command prompt, type "MGRAM" followed by Return. A menu of options will be displayed. The instructions are self-explanatory, and you are prompted for information all along the way. If your address is several lines long, you may enter the extra lines at the colon prompt below the "Company:" prompt. If you do not need the extra lines, you can simply press Return to get to the "Street:" prompt. After each address is entered, you are given a chance to make corrections or enter additional addresses.

When prompted to enter your message, you have the option of uploading a prepared text file or of typing in the information. A carriage return entered on a new line is interpreted as the end of message signal. If you want to include a blank line, press the Space Bar before pressing Return. When your message has been entered, you are prompted for your return address. Unless you have prepared a text file containing this information, you will need to type in your address at the prompts. You may have a confirmation copy sent to yourself if you wish, but you will be charged for the copy. To save time, you may want to bypass the Mailgram menu. The following commands may be typed at the Command prompt:

MGRAM OVER	Provides an overview of general information.
MGRAM INSTRUCT	Provides instructions for using the service.
MGRAM RATES	Displays current rates.
MGRAM SEND	Send a Mailgram.

LOGOFF PROCEDURE

To logoff The Source, get to the Command prompt by typing "Quit." At the Command prompt, type "OFF" and press Return. After a few seconds, a message will be displayed telling you that you are being disconnected and informing you of the amount of time you were online.

CONCLUSION

SourceMail is one of the easiest electronic mail systems to use, and the many sending options make it fast, economical, and versatile. Though the 78-character line length limit may make it difficult to transmit certain types of data, SourceMail allows you to send letters of virtually unlimited length. The mailing list options make it an economical medium for sending messages to a large number of subscribers. However, electronic messages can only be sent to persons having a Source ID, and subscribers are charged for all online time, including reading messages. Access to The Source is available in many overseas countries, making it a relatively fast and inexpensive way to communicate with various international locations.

EasyPlex

EasyPlex
CompuServe Information Service
5000 Arlington Center Boulevard
Columbus, OH 43220
800-848-8199

CompuServe, Inc., a subsidiary of H&R Block, is one of the oldest information utilities online today. CompuServe began as a timesharing service in 1972, but in 1979, in an effort to put its idle computers to work during non-business hours, it introduced MicroNET. The CompuServe Information Division, composed of the Videotex and MicroNET databases, was created in 1980 and is now known as the CompuServe Information Service (CIS).

The EasyPlex electronic-mail system is CompuServe's entry in the instant mail sweepstakes. It's billed by the CompuServe advertising people as "the easiest to use and most readily available" of all the computer-to-computer electronic-mail communications systems. EasyPlex is CompuServe's second-generation electronic mail system, having replaced the company's initial electronic mail service called "Email."

Not only does EasyPlex allow you to correspond with over a quarter of a million CompuServe subscribers, messages can also be sent to users of InfoPlex, CompuServe's electronic mail service designed for business users, and to MCI Mail subscribers.

In addition to email services, CompuServe also offers news, weather and sports; a reference library; home shopping and banking; forums, games, and travel; business and financial services, and much more.

EasyPlex

HOW TO SUBSCRIBE

Subscribing to the service is as simple as picking up a starter kit from your nearest computer or electronics store, or ordering a kit directly from CompuServe. Sign-on is completed by logging on to the service and answering a few questions online. The kits are available for $39.95, and include an attractively-bound User's Guide, a temporary sign-up ID and password, and five hours of free online time that will let you "test-drive" the system. Radio Shack offers a Universal Sign-Up Kit that retails for $19.95. Only one hour free online time is available with this kit. CompuServe subscriptions are also available with the purchase of some computer peripherals. If you receive a subscription this way, all you need do to get full benefit from the EasyPlex service is to write to CompuServe online and order a Users Guide.

RATES

Rates are based on your local time at the point of CompuServe network connection. Some baud rates (450 and 2400) are not available from all locations; while others (4800 and 9600) require hardwire network connections. Connect time rates do not include communication surcharges. To obtain a complete list of these charges, type "BIL-14" at most prompts.

Service costs may be billed to your VISA, MasterCard, or American Express card, or you may request "CHECKFREE" billing, which will allow CompuServe to take the funds from your checking account. For the latter service, you will need to provide CompuServe with your bank's name and address, your checking account number, and the bank's routing transit number. There is no service charge for this option.

Daytime rates apply Monday through Friday, 8 a.m. to 6 p.m. Connect rates are: up to 300 baud, $12.50; 450 baud, $13.25; 1200 baud, $15.00; 2400 baud, $22.50; 4800 baud, $32.50; and 9600 baud, $47.50.

Evening rates apply Monday through Friday, 6 p.m to 5 a.m and all day on weekends and announced CompuServe holidays. Connect rates are: up to 300 baud, $6.00; 450 baud, $7.25; 1200 baud, $12.50; 2400 baud, $19.00; 4800 baud, $29.00; and 9600 baud, $44.00.

Service between 5 a.m. and 8 a.m. weekdays is on an as-available basis, since this time is used for system maintenance, and is billed at evening service connect rates.

The Electric Mailbox

Certain services, such as travel, shopping, and some financial services, are offered at higher rates. There is no charge for the first 128,000 characters of storage. Files are stored 30 days from last access. Additional storage of 64,000 characters is available for $4.00 per week.

ACCESSING THE NETWORK

CompuServe may be accessed through Tymnet, Telenet, Datapac, or CompuServe's own network. There is a communications surcharge made for these services, although CompuServe's own network is by far the most economical. It can be accessed from most major cities. Refer to the appendices for network access numbers.

CONNECTION PROCEDURES

The procedure for connecting to CompuServe depends on the network you are using. Before you sign-on, you will need your ID number and password. Once connection is made, proceed to the logon precedures in the next section.

CompuServe. Dial your local CompuServe network number. When you hear the continuous tone, press CTRL C. If you receive the "HOST NAME" prompt, type "CIS" and press Return. If HOST NAME is not displayed, you will be prompted for your User ID and can proceed to logon.

Tymnet. Dial the Tymnet number for your area and wait for the continuous tone. At the "PLEASE TYPE YOUR TERMINAL IDENTIFIER:" prompt, type "A," but do not press Return. The "PLEASE LOG IN:" prompt is displayed. At the prompt, type one of the following host names: "CIS02", "CIS03," "CIS04," or "CPS01" and press Return. When using a half duplex terminal, you must press CTRL H before entering the host name. If you make a mistake in your entry, simply press Escape and re-enter the host name. You are now ready to logon.

Telenet. Dial your local Telenet number, and at the carrier tone, press Return twice. The "TERMINAL=" prompt will be displayed. Type "D1" and press Return. At the "@" symbol, type "C 202202" or "C 614227," press Return, and proceed to logon. (See the appendices for additional Telenet terminal codes.)

Datapac. Dial your local Datapac number. When you get the carrier tone, enter the appropriate service request signal. For 300 baud, type one period (".") and press Return; and for 1200 baud, type two periods ("..") and press Return. Next, you will be asked to enter a "call request" to indicate

EasyPlex

whether you want to connect via Tymnet or Telenet. Type one of the following codes depending on which service you want to use.

Tymnet

P 1 3106,CIS02
P 1 3106,CIS03
P 1 3106,CIS04
P 1 3106,CPS01

Telenet

1311020200202
1311061400227

Press Return. If Datapac prompts with "HOST NAME," respond by typing "CIS02," "CIS03," "CIS04," or "CPS01," and press Return. If you are using half duplex, press CTRL H before typing the host name. You may now logon.

LOGON PROCEDURES

If your connection is successful, you will be prompted for your User ID and password. Type your ID and press Return, and at the "Password:" prompt, type your password and press Return again. For security reasons, your password will not print on the screen.

   User ID: 79965,565 (Return)
   Password: SPRING/RAIN (Return)

If you have logged on correctly, you will see the date and time of your current session, your last logon session date, and the copyright notice. You are now ready to begin your session.

USING COMPUSERVE

CompuServe Prompts and Commands

EasyPlex offers a choice of three modes of operation: menu, prompt, and command. As a beginner, you will probably want to use the heavily menu-driven beginner's mode. But as you become more comfortable with this mode, try the faster Prompt mode, and finally, experiment with the expert Command mode. (See the Shortcuts section below.) The Command prompt for EasyPlex is an exclamation point ("!"). Be sure to end every command by pressing Return. This tells EasyPlex that you are ready to proceed.

The Electric Mailbox

Getting Help

Help is available online by typing "HELP" or "?" at most "!" prompts. You will be presented with a menu of information available for the particular section of the service you are currently in. Often you will find commands in the help lists that are not available elsewhere. "HELP" followed by a Command name, for example "HELP SEND," will provide specific information about that command.

CompuServe also provides an online Feedback option. If you cannot obtain the help needed by using the Help command, type "GO FEEDBACK" at a command prompt and send a message to the service asking for more assistance. You are not charged for the time you use Feedback except for communications surcharges. And, of course, you can always call Customer Service by dialing 800-848-8990 from outside Ohio and within the contiguous US, or 614-457-8650 from Ohio or outside the contiguous US. They are sometimes hard to reach, but keep trying.

Special Control Keys

The following control characters are recognized by CompuServe.

CTRL A	Causes the current line to finish. The remainder of the information is temporarily stopped.
CTRL S	Halts display at the current character.
CTRL C	Cancels execution of current command or breaks out of the program you are using. This is like a "panic button" that should only be used if you cannot get the system to respond to you commands.
CTRL H	Backspaces.
CTRL O	Stops display of information but does not allow you to resume the display.
CTRL P	Interrupts activity, and usually takes you to the next prompt.
CTRL Q	Resumes display after Ctrl A or Ctrl S are pressed.
CTRL U	Discontinues a line you are typing.
CTRL V	Redisplays a partial line when entered in the middle of typing that line. You may then continue typing the same line after it is redisplayed.

USING EASYPLEX

With EasyPlex, CompuServe simplified the complex command structure of the old EMAIL system into simple, menu-driven electronic-mail commands

EasyPlex

that even a first-time user can handle readily to instantly send and receive electronic mail messages. There are several ways you can reach the email function.

Each time you sign on the CompuServe system, you will see a preliminary menu followed by the CompuServe Main Menu. You can always reach the Main Menu by typing "T" (Top) from any menu system's prompt. (To reach the previous menu, type "M" from any prompt.) From the Main Menu, select "3. Communications/Bulletin Bds." This takes you to another menu titled, strangely enough, "Communications/Bulletin Boards." Select "1. EasyPlex Electronic Mail" to reach the EasyPlex Main Menu.

Whew! That's the long way around. Fortunately, there is an easier way. Just type "GO EASYPLEX" or "GO EASY" at the first system prompt you see, and *voila*, the EasyPlex Menu. Or you could type "FIND MAIL" and get the same result.

There are six options available from the Mail Menu. You may read mail, compose a message, upload a message, use a file from your personal file storage area on CompuServe, consult your address book, or use the Set option. We will discuss each of these in this chapter. Each option in the menu is preceded by a number. To select an option, just pick a number, type it at the Enter Choice prompt, and press Return.

Since the CompuServe User's Guide does not provide definitions of the EasyPlex commands, you might want to type "HELP" at the EasyPlex Main Menu and download the available documentation. Ok, lets get started learning how easy it is to use EasyPlex by sending a letter.

SENDING A LETTER

To create a letter, select "2. Compose A New Message," at the prompt. Your screen will display the caption "EasyPlex Compose" and then prompt you with a "1:," the line marker, as a signal that you can begin to enter text. You can type up to 80 characters per line, and messages are limited to 8000 characters.

The system does not have word wrap so you must press Return at the end of each line. To create a blank line, press Return immediately following a line marker. When you complete the message, type "/EXIT" on a line by itself.

The Electric Mailbox

EasyPlex Compose

Enter message. (/Exit when done)

1:Dear Joseph and Winston,
2:
3:Will meet you in Yalta February 4.
4:Let's make plans for the end of this affair
5:and talk about what we're all going to do next.
6:I'll bring Falla with me to liven up the talks.
7:Eleanor sends her best.
8:
9:Regards,
10:Franklin
11:/EXIT

The /Exit command following your letter will generate the EasyPlex Send Menu from which you may choose to send, edit, type, or file a draft copy of your message. For now, let's send it. Select "1. Send." This generates the "Send To:" prompt. Type the User ID of your correspondent and press Return. Or you may type the name of the recipient if you have stored it in your Address Book. We'll talk about address books shortly. The same message may be sent to up to 10 people, but if sending to more than one individual, separate each ID or name with a semicolon (";"). There is a 10 cent charge for each additional recipient.

Send To:Jack;Mary;75366,565;Steve

Next, you are prompted for a subject for the message. It can contain up to 32 characters. Enter the subject and press Return. You will be queried for your name and then asked if you would like to save this name as your EasyPlex name so you do not have to type it in the future. If you answer "Y," the name or ID you entered will be added automatically to every message you send. And finally, the system displays the address information you have entered and asks if the information is correct. A "Y" response speeds your message on its way. An "N" response gives you an opportunity to correct the address before it is sent.

You may also send a "Receipt Requested" message. When your message is read by the recipient, you automatically receive a confirmation in your mailbox indicating the date and time it was read. To send a receipt requested message, type "SEND/RECEIPT" (SEN/REC) at the Send prompt. There is a 25 cent charge for each receipt.

EasyPlex

Messages can be edited before sending by selecting "2. Edit," from the Send Menu. This will take you to the EasyPlex Edit Menu, where you will be able to specify the lines you wish to edit. Type "?" for an explanation of the editing commands available. (For more advanced users, CompuServe offers an alternate editing system called "Filge," which can also be used in your personal file area to edit files. CompuServe now calls this editor "EDIT," or a non-line-numbered editor. To use this editor, select "Set Options" from the EasyPlex Main Menu.)

To file your message in your personal file area, so that you may later refer to it or send it again, select "3. File Draft Copy." You will be prompted for a filename. Later, we'll learn how to work with these files.

Choose option 4, "Type," from the Send Menu to type the contents of the current message in your workspace. This is a helpful command since it permits you to reread a long message from the top.

UPLOADING FILES TO COMPUSERVE

Composing your letters online is very time-consuming when you have much to say. And remember, you are charged for connect time by the CompuServe people. So, there are times when you might prefer to compose your message offline. It's then that you want to use your word processor and take full advantage of its editing features, and then upload your files to CompuServe. When you save your text file, remember to include a carriage return at the end of each line, and do not allow any line to exceed 80 characters.

To upload a file, return to the EasyPlex Main Menu and type "3. Upload A Message." If you are using CompuServe's own VIDTEX software, the transfer will take place automatically in a special protocol called "B" protocol. If you are using another communications program, transfer will take place in whatever protocol you choose, either XMODEM (or Modem 7) or a capture buffer protocol. For most text files, if your system does not support XMODEM, you may select either "DC2/DC4 protocol" or "No protocol" from the menu given. You are prompted to send the file, in which case you then follow procedures required by your software to upload the file. Remember, you are limited to no more than 8,000 characters in the file, and you need to press CTRL Z to indicate that the transfer is complete if you are using a capture buffer protocol.

When the transfer is complete, return to the Send Menu from which you may choose to send, edit, or type the message.

The Electric Mailbox

USING YOUR ADDRESS BOOK

Mailing lists cannot be prepared on CompuServe, but the service does offer an alternative, the Address Book. The address book can store up to 50 names and ID numbers of the correspondents with whom you frequently communicate. Once the name is entered in your address book, all you need do is enter the intended recipient's name instead of his ID number when you are sending a message. EasyPlex will automatically add the person's ID number from its files, saving you from having to constantly remember ID numbers. Additionally, if you enter your own name in the address book, the system will not prompt you for "From" information.

To enter addresses in your address book, choose "5. Address Book," from the EasyPlex Main Menu or type "ADDRESS" at any EasyPlex prompt. The EasyPlex Address Book Menu is displayed, from which you may choose to insert, change, or delete an entry, list the contents of the book, or enter or change your name. To enter a name, select "1. Insert an Entry," from the menu. You will be prompted for each address, including User ID and name. After you have made all your entries, press Return at a Name prompt. Then, to use the name, simply type the name at the "Name or User ID" prompt when sending a letter.

FILES

Each user has a personal storage area for helping to manage online business. This area is referred to as your "Personal (PER) Area and can be used to accomplish a variety of tasks. You can use your personal area to store letters, drafts, or messages. When you tell EasyPlex to save a message, it is stored in a file in your PER area.

To send an existing file to a correspondent, select "4. Use a file from the PER area," from the EasyPlex Main Menu. You will need to know the name of the file, which can be up to six characters in length.

Once messages are stored in your PER area, you may read, change, or otherwise manipulate them according to your needs by going to the PER area. And, you can create your own file directly in you PER area to be sent by EasyPlex. To reach the PER area, type "PER" at any prompt. This will give you the File Management Menu from which you can view a brief catalog of files; request a detailed directory of files; create and edit files; type a file; delete, rename, or copy a file; and upload or download files. Review your documentation for further details. (See the Shortcuts section below for another way to access the PER area.)

READING MAIL

When you first logon, CompuServe will indicate whether you have electronic mail waiting. This will be your cue to go to the EasyPlex menu. To read your mail, select "1. Read Mail," from the EasyPlex Menu. This generates the the Read Menu. This menu lists the name of the sender followed by the subject of each message.

    1 Bill/Welcome to CompuServe
    2 Bryan/Notes for Next Week's Meeting
    3 Susan/Comments on Digby Draft

    0 READ ALL 3 messages

Note that each of the messages is preceded by a number. To read all messages, type "0." To read any individual message, type the appropriate number. You can also select messages by entering a range of message numbers. For example, "2-3" will only the second and third messages.

Although the selection of messages may not seem important to you, think about how useful the feature could be if you have a dozen messages, but only a few minutes to read them. This feature will allow you to select the most important messages and ignore others until you have more time.

Once you've read a message, EasyPlex asks you how you want to dispose of it. You are taken to the EasyPlex Action Menu where the following options are available.

Delete. Removes the message from your mailbox. It is important to delete messages in your mailbox because, like a physical mailbox, your electronic mailbox has a limit to the number of messages it can hold. Only 20 messages at a time are permitted. The 21st message will be returned to the sender.

File. Stores messages in your PER Area. You are prompted for a filename, which can be up to six characters.

Forward. Sends the message to another correspondent. You may add notes of explanation before it is sent.

Reread. Redisplays a just-read message.

The Electric Mailbox

Reply. Allows you to type a quick reply to the just-read message. You will be placed in Compose mode to type the reply. The ID of your correspondent is entered automatically.

Save. Saves the message in your mailbox.

In addition to these options, there are several others that are not shown on the menu. Simply enter the command at the action prompt.

Scan. Displays all details about mail waiting in your mailbox, including the date and time the letter was posted, the subject, the originator by name and ID number, and the length of the message (in characters).

Read All. Permits you to read all your messages at once. Each message will conclude with the following prompt:

> Last page. Key command
> or <Enter> to continue !

Press Return to go to the EasyPlex Action Menu where you can dispose of the message. If you type "NEXT" at the end of each message, you move on to the next message without interruption.

Undelete. Restores a deleted message. This feature is useful if you accidentally delete a message. The command only works if it is used during the current session. Once you exit from EasyPlex, there is no way to recapture a deleted message.

SHORTCUTS

When you become comfortable with using CompuServe, you may tell the system the types of prompts you wish to receive and the format of your terminal screen. To do this, type "GO DEFALT" at a command ("!") prompt. You will receive a menu allowing you to set various defaults, including the line length in characters for your terminal and which actions you wish CompuServe to take when you logon.

You have the choice of having CompuServe notify you at logon that "you have electronic mail waiting" or of going directly to EasyPlex if you have mail. You may also set up a custom menu, allowing you to select the services you use most by entering a single number. You may wish to add EasyPlex to your custom menu.

EasyPlex

Once in EasyPlex, you may choose from Menu, Prompt, or Command modes, the latter of which is the fastest and most convenient for advanced EasyPlex users, by selecting "Set Options" from the EasyPlex Main Menu.

When reading messages, you may type "Delete" at the "Enter to continue" prompt to delete the message immediately and move on to the next message. This avoids having to use the Action Menu for each letter. It also avoids having to read an entire letter if you can tell from the first page that you are not interested in it.

When you need to access your personal file (PER) area, you may wish to use the command mode. Type "Exit" at a CompuServe command prompt ("!") to get to the PER area "OK" command prompt. At this prompt you may enter Filge editing commands or issue CompuServe commands. Note that at the "OK" prompt, you type "R" to run a program, so you would type "R EASYPLEX" to get back to EasyPlex. Type DIR (directory) or CAT (catalog) to see a list of your files. This avoids having to use the File Management Menu.

SENDING MESSAGES TO MCI MAIL SUBSCRIBERS

An exciting new feature available to EasyPlex users is the ability to send messages to and receive messages from MCI Mail subscribers. This gives both EasyPlex and MCI Mail users access to over a half-million correspondents. Before a message can be sent, you need to know the MCI Mail user ID of your correspondent. However, if it is not known, you may use the subscriber's name just as it is registered in the MCI directory. The preferred method, of course, is the user ID, since proper delivery cannot be assured when using the individual's name. This is because other subscribers may have the same name.

To send an MCI message, at the "Send To:" prompt, type ">MCIMAIL:" followed by the ID.

>MCIMAIL:123-4567

Be sure to include the "greater-than" sign (>) as the first character since this tells EasyPlex that the message is destined for an address outside the CompuServe system. If you are using your correspondent's registered name, type ">MCIMAIL:" followed by the name.

>MCIMAIL:JSMITH

The Electric Mailbox

MCI ID numbers and registered names can be included in your address book. For information on sending MCI Mail messages to EasyPlex users, refer to the MCI Mail chapter of this book.

LOGGING OFF

To logoff the CompuServe system, type "BYE" or "OFF" at any prompt. The "LOG" command also allows you to exit the service, but you are not dropped from the network; instead, you receive a new User ID prompt.

CONCLUSION

CompuServe's EasyPlex system represents a good solid value, a true electronic mail service for the masses. It would be of greater value if the maximum length of messages was increased; 8000 characters only represents about four pages. If you need to send a longer message or upload a lengthy document, you're out of luck, unless you want to break your file up into small chunks. (For example, we were unable to transmit entire chapters of this book to each other via EasyPlex because of this message length limitation.) Also, many of the special sending options available on other email services are unavailable with EasyPlex. However, if your interests lie in an email service that's simple to learn for the casual user, EasyPlex is a system worth investigating. It lives up to its name...Easy.

Delphi

Delphi
General Videotex Corporation
3 Blackstone Street
Cambridge, MA 02139
617-419-3393

Delphi started operations in early 1983 and is now considered one of the big three general-purpose information utilities, ranking with The Source and CompuServe. In addition to its electronic mail feature, Delphi also offers special interest groups for users of most major computers, bulletin boards, conferencing, financial services, games, news, online shopping, a library, and typesetting. Access to Dialog's databases is available as a premium service.

Delphi allows users to send electronic mail messages to subscribers of CompuServe and The Source as well as to other Delphi subscribers. At this writing, this communication link works only in one direction; other services cannot send electronic mail to Delphi. Another nice feature of this system is that subscribers can have letters and other documents translated into a number of languages using the GlobaLink Translation services. MCI Mail may be considered as a future enhancement.

HOW TO SUBSCRIBE

You may make application for membership by calling the hotline at 800-544-4005, or by completing the application blank found in starter kits that are available in many computer stores where they retail for $29.95. The handbook can be ordered online or by calling the hotline. The cost is $19.95 plus $2 shipping and handling.

The Electric Mailbox

All charges are billed to VISA, MasterCard or American Express. Credit card customers have charges applied two or three times a month depending on usage. Direct billing is available for a fee of $3.50 per month. Some services have additional charges.

RATES

Daytime rates apply Monday through Friday, 7 a.m to 6 p.m. your local time zone. Alaska, Hawaii and Guam are handled as Pacific Standard Time, while international zones are handled as Eastern Standard Time. From the continental US, via Tymnet or Uninet, the daytime rate is $17.40 per hour, and $20.40 per hour, via Tymnet, from Alaska, Hawaii and Guam; $18.00 per hour, via Tymnet, from Puerto Rico; $18.00 per hour, via Datapac, from Canada; and $8.00 per hour from international zones other than Canada.

Evening rates apply, your local time, Monday through Friday, 6 p.m to 8 a.m and all day weekends and holidays. Delphi recognizes New Years Day, July 4th, Labor Day, Thanksgiving, and Christmas as holidays. From the continental US, via Tymnet or Uninet, the evening rate is $7.20 per hour; $10.20 per hour, via Tymnet, from Alaska, Hawaii and Guam; $18.00 per hour, via Tymnet, from Puerto Rico; $18.00 per hour, via Datapac, from Canada; and $8.00 per hour from international zones other than Canada.

The above rates apply to all baud rates, 300 through 2400. International connect charges do not include telecommunications surcharges, which are billed separately.

Delphi offers a special local version of the service to subscribers in the Boston area at a reduced rate. Contact the customer service department for more information on this service.

Storage for the first 25,600 characters is free. A charge of 16 cents is made for storage of each additional 1024 characters (2 blocks).

There is no monthly minimum charge.

ACCESSING THE NETWORK

Access to Delphi may be made through Uninet, Tymnet, Datapac (Canada), or a local access number. Phone numbers for these networks are located in the appendices.

Delphi

CONNECTION PROCEDURES

The exact connect procedure depends on the network you are using. When connection to Delphi is made, "Host: call connected" appears on your screen. You can then logon following the procedures in the "Logon Procedures" section.

Uninet. For Uninet access, dial your local Uninet number and wait for the connection. At the "X" prompt, press Return, type a period ("."), then press Return again. To complete the connection, type "DELPHI" at the "SERVICE" prompt. You are now ready to logon.

Tymnet. For Tymnet access, dial your local Tymnet number. When "please type your terminal identifier" (300 baud rate) or a series of random characters (1200 baud rate) appears, type the letter "A." At the "please log in" prompt, type "DELPHI" and press Return. You may now logon.

Datapac. For Datapac access, dial your local Datapac number. Enter one period (".") for 300 baud or two periods ("..") for 1200 baud. Type "Set 2:1, 3:126" for full duplex allowing deletes. Then type "p 1 3106, DELPHI" and press Return. You may now logon.

Local Access. For local access in the Boston area, dial 617-576-0862. When you have a carrier (connect) press Return once or twice and proceed to logon.

LOGON PROCEDURES

Before logging on to Delphi, you need a username and a password. This information is supplied at the time you make application for membership. The time for your Delphi session begins with the time shown when you log in. For those using Tymnet, time is rounded to the next minute, with a 2-minute minimum. Direct dial subscribers have a 1-minute minimum.

Once connected, you are prompted for your username and password. Type them at the appropriate prompt and press Return after each entry. For security reasons, your password will not appear on the screen. The following is an example of a typical logon entry.

    Username: DELPHINE (Return)
    Password: XYXYXY (Return)

If you have logged onto the system properly, you will be informed if you have mail waiting and then greeted by a friendly "Hello" followed by your

The Electric Mailbox

username. The date and time of your logon and the date and time of your last logon are then displayed, followed by the Main Menu.

USING DELPHI

When you logon to Delphi for the first time, you are greeted by a very friendly but fictional character by the name of Max. Max takes you on a tour of the system, helps you change your password in the "security office," takes you to the "terminal room" to identify your terminal type, then on to the "traffic department" to teach you some of the more important traffic signals that Delphi responds to. Here are some of the more important things you need to know.

Delphi Commands

In Delphi, you can type an entire command, e.g., "Mail," or type its first letter or enough of its first letters to make it unique from all other choices. For example, "M" for "Mail." Delphi does not provide a list of abbreviations, so you must learn them by trial and error.

Delphi Prompts

Delphi provides three prompt modes: Brief, Verbose, and Menu. The default value is Menu. Once you become knowledgeable of how the system works, you may find the menus bothersome and time-consuming and want to receive shorter prompts. Shortly, we'll learn how to do this.

The Menu prompt presents a full menu each time you exit to the menu level. For example, the Mail Menu looks like this:

```
MAIL Menu:

BATCH Mailthru              MAIL (Electronic)
CATALOG of Mail Files       SCAN for New Messages
EXIT                        TELEX/Easylink
HELP                        Workspace
GLOBALINK Translation       SetMail

    DMAIL>(Global, Mail, Telex)
```

The Verbose prompt summarizes choices in a one-word prompt. For example, if you use the Verbose option, the prompt looks like this:

    DMAIL>(Ecom, Global, Mail, Telex):

Delphi

And finally, the Brief prompt generates only the prompt sign for the area of Delphi you select. The Brief prompt for the Mail menu looks like this:

    DMAIL>

Some prompts are always brief. For example, "MAIL>" and "WS>" (Workspace). Regardless of the prompt mode you choose, you can always obtain help by typing "HELP" at any prompt.

To choose the prompt mode you want to use, select "Using DELPHI" from the Main Menu, then choose "Settings (PROFILE)" from the Using Delphi Menu, and "PROMPT Mode" from the Settings Menu. Enter the setting you prefer and type "EXIT" to return to the Using Delphi Menu. Now whenever you logon, your prompt mode will be the one selected. To return to the Main Menu, type "EXIT" again.

Getting Help

You can type "HELP" at any prompt to receive help online. You can also receive help by entering a question mark (?) to display a full menu and then a second question mark to put you in the help mode. Each menu also provides a Help option. Type CTRL Z to bring you back to the menu.

Special Control Keys

Here is a list of special control characters you should be familiar with to make your online sessions more enjoyable.

CTRL Z	End input or exit to next higher level
CTRL S	Suspend sending (pause)
CTRL Q	Resume sending
CTRL O	Skip to the end of a file or message
CTRL U	Cancels input for current line
CTRL R	Redisplay current line
CTRL X	Cancel everything typed ahead but unsent
CTRL C	Cancel current activity and start over

Following your tour, you will probably want to wander off and do some exploring on your own. When you have completed your journey, return to this chapter to learn how to use Delphi's electronic mail system.

Before you get started in the Mail function, you should be aware that Delphi's Conference function permits you to conduct online chats with other members. When you are paged, you receive the message, "Do you

The Electric Mailbox

want to talk to (username) (Y/N)?" If you want to chat, consult the Delphi Handbook for procedures on how to do this. But you will probably find that if you are paged while deep in thought preparing an important letter or document, this interruption can be most annoying. To prevent being paged, type "/GAG" from most prompts, including the Main Menu. You will be spared the inconvenience until you choose to re-enable the function by typing "/NOGAG."

USING DELPHI MAIL

Now, let's get down to the business at hand: electronic mail. Select Mail at the Main prompt to access a variety of mail features. You can read, send, reply to, and forward mail to other Delphi users; send mail to users of The Source and CompuServe; send Telex messages to almost any place in the world; or have your mail translated into another language.

The Mail Menu presents the options available in the Mail Utility and is followed by the DMail prompt. Only the types of mail that can be sent are listed in the options following the prompt: Global, Mail, and Telex. But you may also select "Scan" to obtain a list of new messages, "Workspace" to create and edit messages, and "Catalog" to view a list of files you have created. For now, type "MAIL" at the prompt.

Once in the Mail system, a number of commands are available that allow you to move around and perform a variety of functions. These commands are not presented in a menu format, but you can obtain a list of them by typeing "HELP" at the Mail prompt. "HELP" followed by a command gives you a detailed description of the command. Most of the commands are straightforward and mean exactly what they say, so you should have no difficulty in learning them. Since the definitions are not available in the Delphi Handbook, you might want to save them to disk or print them out for future reference.

Let's examine some of your choices and learn how to use them appropriately. Probably one of the first things you will want to do when you enter the Mail utility is to learn how to send a letter.

SENDING A LETTER

In Delphi Mail, the Send and Mail commands can be used interchangeably to send a message to another user on Delphi. Thus, whenever you see the "Send" command in this chapter, remember that it can be replaced with the command "Mail." The same options are available for both com-

Delphi

mands. Do not use these commands to send mail to a correspondent on another system such as CompuServe. We'll learn how to do that later.

To send a letter, type "SEND" at the Mail prompt and press Return. At the "To:" prompt, type the member's username and press Return again. When the "Subj:" prompt is generated, type the subject of your message followed by a Return. You will not receive a "From:" prompt since the system automatically enters this from its files when the message is sent.

Next, you will be asked to "Enter your message below." Type the message, or upload your previously typed file, ending each line with a Return. Text on each line must not exceed 255 characters. Delphi wordwraps, but after typing 255 characters without pressing Return, the message is lost; you are returned to the Mail prompt so you can begin again.

To end the message, type CTRL Z on a line by itself and press Return. If you decide that you do not want to send the message, type CTRL C to Quit and you will be returned to the Mail prompt.

Several editing functions are available when using the Send command, but once the Return key is pressed at the end of a line, there is no way to go back to edit the command line. You can use your Backspace or Erase key to move back to a mistake, enter the correction, then type the rest of the line. Or type a CTRL U to delete the entire line. And at any time you decide you do not want to continue, simply type CTRL C to return to the Mail prompt. The Send screen looks like this:

    MAIL> SEND

    To: Margaret
    Subj: GWTW
    Enter your message below.

    Press CTRL/Z when complete, CTRL/C to quit:

    Thank you for submitting your manuscript. However, our
    editorial staff has decided that there is currently no
    market for this work. Frankly, Miss Mitchell, no one gives
    a damn about Civil War stories.

    CTRL Z

    MAIL>

The Electric Mailbox

Send Options

Delphi does not offer options for sending courtesy or blind copies of letters, nor does it make provision for special delivery of mail or requesting acknowledgement of mail received by an addressee. There are, however, several Send options available that you may find helpful.

Send/Edit. Allows you to use the Delphi Editor to compose messages before sending them. The Delphi Handbook does not list editing commands, but a complete list can be obtained by typing "WORKSPACE" at the DMail prompt. Then at the WS prompt type "EDIT." At the Edit prompt (an asterisk), type "HELP."

Send/Last. Uses the text of the last sent message as the text for the current one. You will be prompted for the username to whom the message is to be sent and then asked for a subject. The text from the previous message is automatically inserted and you are returned to the Mail prompt.

Send/Self. Sends a copy of the message to yourself.

Send/Subject. Permits you to enter the subject of your message on the Mail prompt line. You are not prompted for a subject following the "To:" prompt. To use this option, type "SEND/SUBJECT" followed by an equal sign (=) and the subject title. Enclose the subject title in quotation marks, and do not leave spaces between the equal sign and first quotation mark. For example: SEND/SUBJECT="HOLIDAY SCHEDULE"

UPLOADING FILES TO DELPHI

Since editing on line can be time-consuming and time online is money, you might want to consider preparing text using your own word processing program and then uploading it to Delphi.

Delphi offers several choices for uploading files. You can either use the UPLOAD command, or you can use the more sophisticated upload facility, XUPLOAD, to transfer files in an error-free way. This command uses the Xmodem protocol and enables you to send ordinary printable text files as well as binary files.

To use the Upload command, type "UPLOAD" at the WS prompt followed by the name of the file you want to send. You are asked if you want to send a linefeed after each line, followed by a prompt to enter data. At this point, return to your communication program and take whatever action is

Delphi

required to initiate the transfer. When the transfer is complete, enter a CTRL Z to return to the WS prompt.

To send a file using the Christensen protocol, type "XUPLOAD" at the WS prompt followed by the filename. You are asked if the file is a text file and then prompted to send the file. Take whatever action your communication software requires to transmit the file. You can return to your Delphi session when the transfer is complete. To abort the transfer, type three CTRL C characters.

DOWNLOADING FILES FROM DELPHI

Files can be transferred, error-free, from Delphi to your microcomputer using the XDOWNLOAD command. Type the command at the WS prompt followed by the name of the file you want to receive. You are informed when you can initiate transfer. Return to your communication program and take whatever action is required to receive the file. You can abort the transfer at any time by typing three CTRL C characters. When the transfer is complete, resume your Delphi session.

USING MAILING LISTS

A message can be delivered to several users at one time by entering the username of each person to whom the message is to be sent following the "To:" prompt. Type each name followed by a comma and a space. For example:

    To: Joyce, Bill, Mary

But if you regularly send mail to the same users, you might consider creating a mailing list so you will not have to type the names each time you send the same message to the same people.

Mail lists, called "distribution" lists by Delphi, are created in the Workspace that is accessed from either the Main or DMail prompt. This is a command you want to become familiar with since it is where you create and edit correspondence and develop lists. To create a mailing list, leave the Mail function and return to the DMail prompt by typing "EXIT" at the Mail prompt. At the DMail prompt, type "WORKSPACE."

The Workspace prompt is "WS>." Since you want to create a file, type "CREATE" at the prompt and press Return. You will be prompted for a name to give the list. The name can be anything you like but should be descriptive of the group to whom the mail will be sent. The filename is al-

The Electric Mailbox

ways followed by a period and the file type "DIS" (for distribution list). All mail lists must end with the .DIS type designator. For example, "SALES.DIS" might be the name of a list for your sales staff.

You are prompted for the usernames for the list. Enter each username on a separate line. If you would like to include a comment line preceding the mail list, do this by typing an exclamation mark (!) as the first character in the comment line followed by the comment. Press CTRL Z when the list is complete, or CTRL C if you decide not to create the file.

Here is an example of a mail list file:

> ! New personnel June 1986 (comment line)
> Margaret
> John
> Betty

To use the mail list, type "SEND" at the Mail prompt, and at the "To:" prompt type the filename you gave to the list. Precede the filename with an @ symbol. Do not enter the ".DIS" file extension. Names of individuals whose names are not on the list can be added by placing their names on the same line as the filename, separated by a comma but no space. The filename may appear in any position in the "To:" field.

> To: @SALES,RANDI,PAUL,VAL,DENNIS

If you realize after sending the message to your mail list that a name was forgotten, you can resend it provided you have not entered a new mail command or left the Mail function. Type "S/L" at the Mail prompt, and at the "To:" prompt type the additional names. Since the S/L command does not copy the subject of the message, you must either re-enter the same subject or provide a new one at the "Subj:" prompt. Then press Return to send the message and return to the Mail prompt.

An interesting feature of the Send command is that it can send an existing file to another user. To do this, at the Mail prompt type "Send" followed by the filename. For example, to send a file with the filename "Smith.Ltr," make the following entry.

> MAIL> SEND SMITH.LTR

You are prompted for the username and subject. The file is automatically inserted and the message sent. The text will not appear on the screen.

Delphi

FILES AND FOLDERS

Before you learn to read messages, let's take a look at Delphi's file system. You must be able to understand this feature in order to move about easily in the Mail Utility and handle messages effectively. The file system consists of three tiers: files, folders, and messages. Files hold folders and folders hold messages.

When you log on to Delphi, you will find that one file has already been established for you. It's called "MAIL.MAI" and is considered your mail file. You can create as many additional files as you wish. The Delphi Handbook and online help feature will tell you how to do this. For now, let's see how to work with the "MAIL.MAI" file. Three default folders exist in this file: NewMail, Mail, and Wastebasket.

As new messages are received, they are placed in the Newmail folder where they remain until you read them. Once a message is read, it is automatically removed from the Newmail folder and placed in the Mail folder, unless it is deleted. Deleted messages are filed in the Wastebasket folder where they remain until you exit mail or use the Purge command. (Use your online Help feature to find out how to use this command.) You can view messages in the Mail or Wastebasket folders by typing "DIRECTORY" at the Mail prompt followed by the name of the appropriate folder. You will learn more about working with files and folders when you review the section on Reading Mail.

READING MAIL

One of the first things you will want to do each time you enter the Mail Menu is to check your files to see if you have any new messages. To do this, type "DIR/NEW" at the DMail prompt. You will see a list of all new messages by name of the person sending them, the subject, and the date and time of receipt. If there is no mail waiting, you receive the message "No New Mail."

To read or take any action on these messages, you must first enter "MAIL" at the DMail prompt.

There are several ways to read mail. You can type "READ" at the Mail prompt to display the oldest message in your Newmail folder, one message at a time. If there are no new messages, the oldest message in the Mail folder is displayed. The Mail prompt appears at the end of each screenful. Just press Return at each prompt to continue reading messages. When a message consists of more than one page, you are

The Electric Mailbox

prompted to "Press RETURN for more..." You are then returned to the Mail prompt where you can press Return to continue reading the current message, enter a new Read command to read a different message, perform some other Mail function, or Exit mail. If you decide to stop reading the message, you can always go back to read it later. If "READ" is typed and there are no more messages, the phrase "No more messages" is displayed.

When a new message is received while you are in the Mail utility, the banner "New mail on Node (Nodename) from (username)" prints on your screen. To read this message immediately, type "READ/NEW" at the Mail prompt. After reading your new mail, you can return to your previous activity.

The name of the mail folder in which a message is filed is printed in the upper right corner of the message, and a message number is printed directly under the Mail prompt. If you want to read a specific message listed in the directory, just type "READ" at the Mail prompt followed by the number of the message. For example, to read message 3, type "READ 3" or just "3" and press Return. Once the message is read, it can be filed in a file and folder of your choice, forwarded to another Delphi user or replied to. These actions can only be taken on the current message, that is, the message you are reading. If you do not want to take an action, simply type another command and the just-read message will be returned to the "MAIL.MAI" file. Let's look at the actions you can take on messages you have read.

Read Options

Delete. Deletes the current message and returns you to the Mail prompt. Now, if you type "DIRECTORY" at the Mail prompt, you will see "(Deleted)" entered next to the message number. Similarly, if you type "WASTEBASKET" at the Mail prompt, you will find that the deleted message appears in the Wastebasket directory.

File/Move. The File and Move commands are used interchangeably. A just-read message is automatically filed in your MAIL.MAI file. This command places a message in a new folder in the file or in a different file and folder, if you identify one. To add the current message to an existing folder in your Mail file, type "FILE" at the Mail prompt followed by the foldername you want to use. For example, to place a message in a file named "statistics," you would type "FILE STATISTICS." To read the filed message, type "READ STATISTICS" at the Mail prompt.

Delphi

Reply/Answer. The Reply and Answer commands are synonymous and may be used interchangeably. Your reply can be a new message, in which case you are prompted for the text. The username and subject are automatically entered from the message to which you are responding. Or you can reply by requesting that an existing file be sent. You are prompted for the username and subject. The file is entered and you are returned to the Mail prompt. Three options are available when using the Reply command.

Reply/Edit. Permits you to edit the reply.

Reply/Last. Permits you to send a reply to a message using the same text as the last message you sent.

Reply/Self. Permits you to reply to the current message and to send a copy of the reply to yourself.

Forward. Forwards the current message to another user.

Next. Takes you to the next message. The message received by entering this command is now considered your current message and can be forwarded, filed, or replied to.

Back. Allows you to return to the message just prior to the current one.

Search. Allows you to locate a particular message. For example, suppose you remember receiving a message containing an important meeting date, but can't remember the message that contained it. You can request a search of your files for messages with the word "meeting" in them. This is called a "search-string." For example, "MAIL> SEARCH MEETING." The search begins with the first message in your files and includes a search of the "To," "From," and "Subj" fields, and the text of the message. If the search-string is found, the message is displayed. If it is not found, the next message is searched for the same string. This process continues until all messages have been searched or the specified search-string is located.

Copy. Copies a message from one folder to another without deleting it from the current folder.

Current. Permits you to read a message again.

First. Displays the first message in the current mail folder.

Last. Displays the last message in the current folder.

The Electric Mailbox

To exit from Mail, type "EXIT" or CTRL Z at the Mail prompt. The Exit command empties the Wastebasket folder, unless you issue the "Set Noauto Purge" command. You can review the Set commands online.

SHORTCUTS

If Delphi is busy sending you information and you know what you want to do next, you can save time by typing the next command before Delphi completes its task. When Delphi finishes, your input will appear on the screen just as it would have if you waited. If you change your mind before the system catches up with you, just type CTRL X. Everything that you had typed will be deleted, and you can start over. Also, remember that most Delphi commands can be abbreviated to one letter or as many letters as are needed to make the command unique.

SENDING A BATCH MAILTHRU MESSAGE

One of the disadvantages of some electronic mail systems today is the inability of users to communicate with subscribers of other systems. Delphi has made a breakthrough by providing a program called Mailthru that gives you the ability to send messages to users of The Source and CompuServe. Unfortunately, the current system only works one way. Messages cannot be received from other systems. But it is a start, and one that is needed badly in all email systems.

As messages are prepare for The Source and CompuServe, they are batched and mailed each day at 6:00 a.m. If for some reason the service is not available, Mailthru attempts to send it the following day. A charge of 75 cents is made for each message sent to another system and the cost is billed to your credit card.

To use Mailthru, type "BATCH" at the DMail prompt. You are asked for the name of the service to which you want to send the message and then prompted for the ID of your correspondent and a subject title. Follow each entry by pressing Return. When you receive the "Enter your message below" prompt, type the message, ending it with a CTRL Z on a line by itself. You are informed when the message is sent, thanked for using the service, and returned to the Mailthru prompt. To send another message, type the name of the service and proceed as above. A CTRL C will abort the message.

Delphi

SENDING A TELEX

Telex messages can be sent by Delphi members through Western Union EasyLink to any of the thousands of telex users throughout the world. The rate for sending a telex message to a US location is 50 cents per 100 characters (or fraction thereof). Messages sent to Canada and Mexico are $1 per 100 characters, and messages sent to other countries are $2 per 100 characters.

To reach the Delphi Telex function, type "TELEX" at the DMail prompt. You are prompted for the telex address or EasyLink ID, and asked if you are sending a domestic or international message. You are then asked whether you are sending to an ITT, RCA, WUI, TRT, FTCC or Graphnet address. Respond to the question. You may now type your message. Each line must be no longer than 67 characters, and the message cannot exceed 50,000 characters. Press CTRL Z to end the message or CTRL C to abort. Delphi provides the necessary codes.

International telex messages are prepared the same as domestic ones, with the exception that you must respond "yes" to the question "Is the Telex address outside of the USA?" You are then prompted for the name of the country to which the message is going and for the international code for that country. A complete list of telex country codes is available in the "Using Delphi" section of the system. To access the list, type "TELEX CODES" at the Using-Delphi prompt. Complete the remainder of the message just like a domestic telex message.

REQUESTING A GLOBALINK TRANSLATION

An unusual feature of Delphi is its GlobaLink Translation service. The service is in operation 24 hours a day, 7 days a week. Correspondence and other written documents can be translated from English into French, German, Italian, Portuguese, and Spanish, and visa versa. Translations into most other languages can be requested, but GlobaLink requires advanced notice of the need.

Turnaround time depends of the complexity of the material, its length, and the time of day GlobaLink receives it. Anything sent after 9:00 p.m. Eastern Standard Time is not considered received until 8:00 a.m. the following day. In general, the following turnaround times apply: for Telex messages up to 1,500 characters, 3 hours maximum; non-technical material up to 12,500 characters, 24 hours maximum; correspondence up to 25,000 characters, 48 hours maximum; and complex or technical material up to

The Electric Mailbox

5,000 characters, 24 hours maximum. (Complex or technical material over 5,000 characters is done on a job-by-job basis.)

A flat rate of 3 cents per character plus a $3.00 handling fee is charged for all transactions and there is a $20.00 minimum per translation. Translation costs are charged to the member's Delphi account.

To access GlobaLink Translation, type "GLOBAL" at the DMail prompt. You are asked for the language from which the message is to be translated and then for the translation language. There are a variety of ways a translated document can be sent. You can send it to another Delphi member or by Western Union Telex, or have a hardcopy prepared and mailed to your correspondent. If you have special instructions for the translator enter them at the "special comments" prompt. Type your message, ending with a CTRL Z on a line by itself. You are prompted for entries each step of the way.

LOGGING OFF

You must exit Delphi from the Main menu. No matter where you are in the system, you can always press CTRL Z or CTRL C to return to a menu. From the current menu, type "EXIT" to return to the Main menu. Then at the Main prompt, type "EXIT" or "BYE" to leave the system. You are informed of your logoff time and the length of the session.

Delphi makes it easy for you to logon again by providing the "please log in:" prompt at the end of the logoff sequence. Just type "DELPHI" and logon in the usual way. Otherwise, clear your buffer, if required, and take whatever action your software requires to exit to your operating system.

CONCLUSION

Delphi is an easy system to learn and to use but it tends to be a little slow. For example, you must always return to a menu in order to move to another part of the system. There are no short cuts. If you are willing to put up with a little sluggishness, Delphi is a great system, well worth considering. Delphi offers a wide range of electronic mail services and its Batch Mailthru and GlobaLink Translation features are unique and valuable enhancements. You should be able to accomplish most of your business or home correspondence tasks using the system's many features.

ECHO

ECHO
4739 Alla Road
Marina del Rey, CA 90291
213-823-8415

ECHO, Electronic Communications for the Home and Office, is a relatively new but rapidly growing telecommunications system. It has been called the David that's flexing its muscles against the Goliaths of electronic communications. ECHO started operations in March 1984 and now boasts a membership of 28,000, including individual and business accounts. Although the primary focus is on electronic mail, ECHO also supports a bulletin board service.

ECHO offers two options to its subscribers. Option A is a flat rate service designed for individual users and provides email services, access to public bulletin boards, an edit function, online storage and utility functions, and discount shopping. Option B is a metered rate service and is used by most subscribers since it is the most economical for long distance access. This service is customized for each client according to the client's membership size, geographical location, and system usage requirements. It provides a private network for correspondence and information dissemination, linking individuals and offices through a nationwide system.

Although option B offers all the functions available in Option A, it is entirely separate and secure from the public ECHO system. The user receives a user directory, proprietary bulletin boards, and conferencing (an optional feature). The private network allows an organization to track use of the system by way of detailed usage and accounting reports, resulting in better management of resources and personnel, and greater productivity.

HOW TO SUBSCRIBE

Make application by calling 213-822-ECHO and requesting a Subscriber's Agreement. Within 48 hours of receipt of your application, ECHO will mail your guide, user ID, password, and access numbers. There is a documentation fee of $25.00, which includes your ID and Systems Guide.

RATES

Daytime rates apply Monday through Friday, 7:00 a.m. to 6:00 p.m. For Option A, charges are $8.00 per hour at 300 baud, and $10.00 per hour at 1200 baud. Rates for Option B are $10.00 per hour for 300 baud, and $12.00 per hour for 1200 baud.

Evening rates apply from 6:00 p.m. to 7:00 a.m. all other times. For Option A, charges are $4.00 at 300 baud, and $8.00 per hour at 1200 baud. Rates for Option B are $5.00 per hour at 300 baud, and $12.00 per hour at 1200 baud.

There is a monthly usage charge of $10.00 for Option A subscribers. No monthly charge is made for Option B members. All rates are calculated using your local time zone.

The first 212,000 characters stored on the system are free. Additional storage may be purchased in blocks of 212,000 characters at twenty cents per day per block.

ACCESSING THE NETWORK

Access to ECHO is made through the BTS (Budget Time-Share) network. Local access telephone numbers are provided at the time you make application for service.

CONNECTION PROCEDURES

Before you logon to ECHO, you will need a network code, an ID number, and a password. To access BTS, dial the number supplied at the time you applied for service. When you are connected, press Return, type a period ("."), and pressing Return again. At the "Service:" prompt, type "ECHO," followed by the network code provided by ECHO. At the "Terminal ID =" prompt, type an "A" if you are using half duplex, or a "B" if you are using full duplex. If your connection is successful, "Welcome to BTS" is displayed. Press Return and you are ready to logon to the system.

ECHO

LOGON PROCEDURES

When you are connected to ECHO, you will be prompted with a period ("."), the ECHO system prompt. Type an "L" (Logon), press the space bar once, type your ID, and press Return. The system responds with the "Enter Password" prompt. When you have entered your password, press Return again.

    .L ECHOXXXX (Return)
    Enter Password: REDROSES (Return)

You receive a banner displaying the time, and date of your logon. That's it. You're now logged on and ready to enjoy your first ECHO experience.

USING ECHO

ECHO is a menu driven electronic mail service that combines menus with explicit on-screen instructional material. Each prompt is followed by the steps required to carry out the chosen command. When options are available, questions are asked to which the user must respond with a "Y" or "N," and sometimes a response must be verified. Although this type of instructional help is particularly useful when you first start using the system, it soon becomes cumbersome and annoying.

Fortunately, ECHO provides a way to bypass these instructions. The instruction level can be set to Off, which results in commands being displayed, but instructions suppressed. This makes your session move along more rapidly.

The ECHO Main Menu offers four options: Communications, Information, Utilities, and ECHO Hot-Line. In addition, a command is also included for logging off the system. The logoff selection is available on all menus. Each menu selection is preceded by a number, and these numbers are used to enter your selection. If you attempt to type the name of your selection, you receive the message "***Error on entry...please respond as indicated***."

Getting Help

Help can be obtained during regular business hours Monday through Friday by calling 213-822-ECHO. Questions can also be entered online using the ECHO Hot-Line. Responses are usually received within 24 hours. Help is always available online by typing "HELP" at any prompt.

The Electric Mailbox

Special Commands

The Main Menu is followed by the system prompt, a period ("."). There are several commands to become familiar with before getting started. These can be used from any system prompt, or any time you are asked to respond to a yes or no question.

H   *Help.* Displays a list of systems commands.

Q   *Quit.* Aborts the present task. You remain in the function you were in before you entered the command. This is a helpful command, for example, if you are creating a message and want to start it over again.

E   *Exit.* Exits the program and return to the menu from which you made your last selection.

M   *Main Menu.* Returns you to the Main Menu.

L   *Logoff.* Ends the session and logs you off the system.

Echo Storage

ECHO does not support a formal filing system where messages and documents can be filed in folders of your choice, but it does offer a generous amount of free disk space (212,000 bytes.) If you find that additional space is required, it can be ordered online and is usually available for use within 24 hours. The storage area is called ECHOdisk, and it is where you file all sent and received messages that you want to save.

With the preliminaries out of the way, you're ready to see just how ECHO's email system works. To get to the email function select "1. Communications," from the Main Menu, and then choose "1. E-Mail" to reach the "E-Mail Functions" Menu. At this menu, you may choose to Send, Read, Erase, perform a group functions or check for waiting email.

SENDING A LETTER

To create new mail, type "1" to Send from the E-Mail Function Menu. This displays the Send Options Menu. Type "1" again at the "." prompt to reach the "Create New E-Mail" option. You are prompted for each entry.

First you are asked to enter the ID of the person or the name of the group (mailing list) to whom the message is to be sent. You'll learn how to create

ECHO

mailing lists (otherwise known as distribution lists) shortly. You also have the option of having your message printed and mailed by ECHO to a system user or to a friend or associate who is a non-ECHO user. For now, let's create a message to send a message to another user.

At the ECHO prompt, type the recipient's ID and press Return. Instructions for entering the text are displayed, followed by a prompt telling you to "Enter Text." Each new line of text is preceded by the system prompt. Press Return at the end of each line. Each line of text can contain no more than 80 characters. If this number is exceeded, the message "***Line length exceeded 80 characters...please start the line again" is displayed. The too-long line will not appear in the completed message. Just retype the line and continue with the text. To leave a space between lines, press Return at the system prompt. You may use your backspace key to make corrections on the line on which you are typing.

>	Enter recipient's ID (or recipient Group name):
.ECHOXXXX
ENTER TEXT:
.George,
.
.Love your new choppers. By the way, what happened
.to the cherry tree?
.
.Martha
.Q

End the message by typing a "Q" (for Quit) on a line by itself and pressing Return. This generates the E-Mail Complete Menu from which you may choose to send the message as is, or elect to edit it. (See "The ECHO Editor" chapter in the ECHO guide for instruction on editing messages.)

To send the message as is, type "1" at the the E-Mail Complete Menu prompt. Next, you are asked to provide a subject line for the message and then to confirm that the subject entry is correct. The subject line may contain up to seven words.

>	Describe E-MAIL in seven words or less:
(for identification to recipient)
(Q to quit)
.PRODUCTION FIGURES
Verify description: PRODUCTION FIGURES ...Okay (Y/N)?
.Y

113

The Electric Mailbox

If you want to change the subject entry, simply answer the prompt by typing "N" and you will receive a new prompt. A "Y" response takes you to the E-Mail Priority Menu from which you can select the method you want to use to send the message.

The E-Mail Priority Menu lists three choices for sending mail. Select option "1" to send the message on an "Urgent" basis with a request for view confirmation. (You are asking for a Return Receipt to indicate that the message has been seen by the addressee.)

Select option "2" to send an urgent message without view confirmation. Both options cause a banner to be printed across the recipient's screen at logon to indicate that urgent mail is waiting. If you chose option 1, once the message is read by the recipient, you receive confirmation in your mailbox that indicates the date and time it was read.

Option 3 on the Priority Menu is called "Regular." This option causes the message to be sent without "urgent" tags. You are informed that it has been sent, then you are asked if you want to save the message on your ECHODisk. To save the message, type "Y" at the prompt. And finally, you are asked if you want to continue sending email. If you answer "Y" you are returned to the Send Option menu. Otherwise, you are returned to the E-Mail Function Menu.

Two additional options are also available from the Send Menu:

Resending Old Mail. Sends a message that was previously received or created and saved on your ECHOdisk.

Modifying Old Mail. Permits you to edit a message that was previously received or created before you send it.

UPLOADING FILES TO ECHO

If you find sending a letter using the "Create New E-Mail" option tedious, you might find it easier, not to mention more cost-effective, to create your letter on your own word processor and upload it to ECHO. ECHO suggests that you consult an ECHO representative before attempting to upload files since all software does not perform faultlessly on ECHO. Your file must contain a carriage return at the end of each line, and it must be capable of waiting to receive a period (.) from the ECHO computer. Unless these requirements are met, ECHO cannot guarantee the successful uploading of files.

ECHO

To upload a file, go to the Send Options Menu and type "4," Upload a file from your PC, at the prompt. At the Period prompt, type the recipient's ID or a mailing list name. You will be instructed to "Begin Upload:" Take whatever action is required by your software to send files. When transmission is complete, type a "Q" to return to the E-Mail Complete Menu. To complete the message, follow the instructions for sending new email.

USING MAILING LISTS

A mailing list, or "Group Function" as ECHO calls it, can be created to prepare lists of persons with whom you correspond on a regular basis. Each list is identified by a group name, so when you want to send the same message to all members of the group, just enter the group name instead of typing each ID separately. ECHO will automatically send the identical message to each name listed under the group name.

To create a group list, select "4. Group Functions," at the E-Mail Function Menu. The Group Function Menu will be displayed. Select "2. Create," from this menu to prepare your distribution list. At the "Enter group name" prompt, type the name you want to give the list. The name may consist of up to eight alphanumeric characters, but must not contain spaces.

    Enter Group name:
    (Q to quit)
    .PENPALS

You are then prompted for the IDs to include on the list. Type each ID on a separate line. When the list is complete, type "Q" to return to the Group Functions Menu.

    ENTER ID's:
    .ECHOX11X
    .ECHOY22Y
    .ECHOZ33Z
    .ECHOX44Y
    .Q

Your list is now ready and can be used at any time you want to send the same message to each name on the list. To use the list, type the group name you have given the list at any "Enter Recipient's ID" prompt. Here are some options available for use with mailing lists:

Displaying A Group List. Displays a list of all groups you have created.

The Electric Mailbox

Modifying A Group List. Adds or removes IDs from a list.

Removing A Group List. Deletes group names. When you delete a group name, the group name and all the ID's associated with it are erased.

FILES AND FOLDERS

Whenever you create a message or receive a message from another user, you are asked if you want to save it on your ECHOdisk. If you respond "Y," the message is assigned a prefix of "SMAIL" (sent mail) or "RMAIL" (received mail), followed by a message number. Numbers begin at "001" and increase each time a new message is saved. For example, "SMAIL008," or "RMAIL015." These numbers are important because they are always used to read saved mail. We will learn how to use them in the "Reading Mail" section of this chapter.

The Erase function allows you to delete old or unwanted messages from your ECHOdisk. It is important to remember that once a message has been deleted, there is no way to retrieve it. To erase a message, type the SMAIL or RMAIL number of the message to be erased at the E-Mail Function prompt. If you are not sure of the number, press Return to view a directory. After you enter the number, a mini-directory is displayed showing the number, date and time the message was received or sent, the ID of the person sending or receiving it, and the subject. You are then asked if you are sure you want to erase it.

READING MAIL

To read mail, select "1. Communications," from the Main Menu, and then select "1. E-Mail," to display the E-Mail Functions Menu. There are two ways you can read mail: the "Check E-Mail Waiting" option, and the "Read" option.

Check E-Mail Waiting

To read new mail, select "5. Check E-Mail Waiting," from the E-Mail Functions Menu. If you have mail waiting, a list of messages is displayed.

ECHO assigns each new message with a system number, which appears as the first item in the checklist and is followed by the ID of the person sending the message, the date sent, and the subject. If you receive an urgent message and the sender has requested view confirmation, an asterisk (*) appears between the systems number and the "From:" entry. Fol-

ECHO

lowing the checklist, you are prompted for the system number (SYS#) of the message you want to read.

The SYS# is a special number assigned by ECHO as a means of identifying a message from among the thousands of messages being routed to and from all ECHO users. Type the SYS# at the ECHO prompt. The message is displayed and is followed by a prompt asking whether you want to save it to your ECHOdisk. If you respond "Y," the message is assigned an RMAIL message number. You are informed of the number assigned to the message and then returned to the E-Mail Function Menu. To read the message again, you must enter the Read function. If you answer "N" the message is deleted. Be aware that once deleted, a message cannot be retrieved.

Read

To read mail that has been saved, choose "2. Read," from the E-Mail Function Menu. You will be prompted for the E-Mail number of the message you want to read. If you do not know the number, press Return to view the E-Mail Directory. ECHO provides three directories: one for messages you have received; one for messages you have sent; and the third for specific E-Mail. Type the number of the directory you want to view and press Return. When the directory is displayed, you will be prompted for the SMAIL or RMAIL number of the message you want to read. Whenever prompted for a message number, you can always press Return to view the directory again.

If you want to search for mail that meets certain criteria, type "3" at the E-Mail Directory prompt to generate the Specific E-Mail Menu. You may then call up a list of letters that were:

- Sent to a Specific ID or Group
- Received from a Specific ID
- Sent During a Specific Month, or
- Received During a Specific Month

SENDING A HARDCOPY

A hard copy of your message can be printed and mailed to an addressee of your choice. The individual does not have to be an ECHO system subscriber. To send a hardcopy, choose "1. Create New E-Mail," from the Send Options Menu. When prompted for the recipient's ID, type "HARDCOPY." You will be prompted for the name, street address and city/state/zip of the addressee. Each entry is verified by the system, so if you make a mistake, it can be corrected without entering the Edit Mode. The prompts for the remainder of the procedure are the same as those for creating any other message. (See the Sending Mail section of this chapter.)

The following rates are charged for the hardcopy service:

Number of pages	Price per page
1 - 249 pages	$1.75
250 - 499	$1.60
500 - 999	$1.50
1000 - 2,499	$1.40
2,500 or more	$1.10

LOGOFF PROCEDURE

Logging off ECHO is simple. Just type "L" from any prompt. You are thanked for using the system, informed of the amount of time you were connected, and the hour and date of logoff.

CONCLUSION

ECHO is a straightforward, no-frills, and low cost electronic mail system that seems best suited for individuals and small businesses. It has a rapidly growing subscribership and will no doubt be adding other enhancements in months and years to come. In fact, the next generation ECHO system will offer an optional service by which ECHO's computer dials out to the recipient's PC and delivers a message directly into the PC without requiring the recipient to logon to the system to retrieve the message. Moreover, the recipient need not be an ECHO subscriber. This promises to be an exciting feature. You can't beat the price of ECHO, and it's well worth the money.

Dialmail

Dialmail
Knowledge Index
Dialog Information Services, Inc.
3460 Hillview Avenue
Palo Alto, California 94304
415-858-3796

Knowledge Index (KI) is an online information service that currently supports 28 databases. It is a simplified version of Dialog Information Services, a larger online information service that offers more than 250 databases in all areas of interest. Because KI is available only during non-peak hours, it is considerably cheaper than Dialog.

Dialmail, KI's email service, was initiated in April 1985. Created by Dialog, Dialmail offers subscribers the ability to send and receive messages electronically with other KI users; create files of text that can be stored, edited or delivered to other users; participate in conferences; create order forms; and use bulletin boards to post or read announcements.

Paper copies of letters and documents can be prepared for delivery from Dialmail to third parties via US Mail. Dialmail is now available to Dialog customers.

HOW TO SUBSCRIBE

Make application by calling or writing Dialog and requesting a customer agreement. An initial registration fee of $35.00 is charged. This includes two hours of free online time, if used within the first 30 days, and the User's Handbook.

RATES

A flat rate of $18.00 per hour is charged for Dialmail services. This includes telecommunications costs. There is no monthly minimum. There is no extra charge for 1200 baud. The hours are Monday through Friday, 6 p.m. to 5 a.m. the next morning; Saturday, 6 p.m. to midnight; and Sunday, 3 p.m. to 5 a.m. the next morning. All times are your local time. Changes in the hours of availability occur only in areas that do not observe Daylight Savings Time.

Online storage is charged at the rate of one-half cent per page (1024 characters) per night.

ACCESSING THE NETWORK

Knowledge Index is accessed through the Tymnet, Dialnet, Uninet, or Telenet telecommunications networks. Local access numbers for these networks are available in most large cities. Refer to the appendices for the telephone numbers. Canadian subscribers may access the telecommunication networks through Datapac. Subscribers who live within the 415 telephone area code may dial direct. However, direct dial access is available only at 300 baud.

CONNECTION PROCEDURES

The procedure for connecting to Knowledge Index depends on the network you are using. When connection is made you receive the message "Dialog Information Services - Please Logon." Then follow the procedures in the "Logon Procedures" section.

Tymnet. For Tymnet access, dial your local Tymnet number and wait for the carrier tone. Type "A" at the "Terminal identifier" prompt. When accessing the network at 1200 baud, the message may be garbled. Wait for it to stop before typing the "A." When Tymnet responds with the "Please log in:" prompt, type "KI" and press Return. You are now connected and can proceed to logon.

Dialnet. For Dialnet access, dial the Dialnet number for your area. After the connection is made, type "A." When Dialnet responds with the "Enter Service" prompt, type "KI," press Return, and proceed to logon.

Uninet. For Uninet access, dial the Uninet number and wait for the connection. At the "X" prompt, press Return, type a period ("."), then press Return again. The Uninet pad and port entry is displayed followed by the

Dialmail

"Service:" prompt. Type "DLG;KI" and press Return. When the logon prompt is displayed, proceed to logon.

Telenet. For Telenet access, dial the local Telenet number. When you hear the high-pitched tone, press Return twice and wait for a response. When Telenet responds with the "Terminal:" prompt, enter your terminal identifier. "D1" is most commonly used for microcomputers. See the appendix for Telenet for other terminal types. When you receive the "@" symbol, type "C 41548K." This number tells Telenet to connect you to Knowledge Index. You are now connected to KI and can proceed to logon.

Datapac. Datapac users must use one of the following addresses when accessing the public telecommunication networks:

```
Uninet:     1 3125 4150005000,KI
Tymnet:     1 3106 900803,KI or 1 3106 900061,KI
Telenet:    1 3110 2130023611 or 1 3110 2130017011
            or 1 3110 4150004811 or 1 3110 4150002011
```

LOGON PROCEDURES

A password and an account number are required in order to logon to KI. These are provided after you register for the service and receive your order confirmation sheet. At the "?" prompt, type your six-character account number (the number that begins with "U"). Press Return. You will then be prompted to enter your password. Type your password and press Return again. Your password does not print on the screen. Your entry will look like this:

```
?U12345 (Return)
?XXXXXXXX (Return)
```

If your entires are correct, Knowledge Index will greet you with brief instructions on how to start using the system. If you have mail waiting, a notice announcing this appears following the welcome banner. You are given the time your accounting starts and instructions for obtaining help. When ready for your input, the system will signal with the Knowledge Index prompt, a question mark ("?").

USING DIALMAIL

Dialmail is menu driven with an option for working with long or short menu formats. The short menu will probably be sufficient once you become familiar with the service and its commands. Long menus are listed vertically

The Electric Mailbox

with brief definitions of use, while short menus are listed horizontally giving only the command itself.

Dialmail Prompts

The Dialmail prompt is "Your Command:," except in the Edit and Input modes where a question mark ("?") is used. Whenever "Your Command:" is seen, you are expected to respond by typing a command and pressing Return.

Getting Help

The Dialmail user's manual, "Dialmail Basics," gives the basic information needed to get started, and online help is available from most menus. To obtain help online, type "HELP" to display all the options, or type "HELP" followed by a command to receive help on a particular command. Customer service can be reached at the following numbers to respond to your questions: US, except California, 7:00 p.m. to 11:00 p.m. EST, 800-334-2564; California, 7:00 to 11:00 p.m. EST, 415-858-3796; Canada, 9:00 a.m. to 6:00 p.m. EST, 416-593-5211; and Canada, 6:00 p.m. to 9:00 p.m. EST, 416-535-1323.

With the preliminaries out of the way, let's get started. To begin your Dialmail adventure, type "BEGIN MAIL" at the "?" prompt.

The primary commands used in Dialmail are Scan, Read, and Create. These commands are used to manipulate messages that are stored in your Inbox, on your desk, and in your file folders. Commands may be entered by using the full command name or simply the first letter (in most cases) of the command. For example, type SCAN or S; CREATE or C; READ or R. As you become familiar with the system, some menus can be bypassed by entering the Main Menu command followed by a space and the sub-menu command. For example, you can type "READ" (from the main menu), then "FOLDER" (from the Read Menu) or simply "R F" from the Main Menu, to go directly to reading your list of folders. Let's get started by learning how to send a letter.

SENDING A LETTER

"Create" is the command used to compose letters. To send a letter, type "CREATE" at the Main Menu. At the "To:" prompt, type the name of your correspondent and press Return. The "To:" prompt repeats until you press a Return following the prompt without typing a name. Only the last name

Dialmail

needs to be entered because Dialmail automatically enters the first name from its files.

The "subject" prompt is then displayed. The subject can consist of up to 40 characters and can contain alphanumeric characters and punctuation. Type the subject and press Return. Next you are prompted for the text. Enter the text, ending it by typing a period (".") on a line by itself and pressing Return. This is your signal to Dialmail that you have completed the message. The Create Menu of options is then displayed. Here is a sample letter.

> To: Shaky
> Message addressed to: Shaky Arms Hotel
> To: (Return)
>
> Subject (up to 40 char): Reservation
>
> Enter text, ending it by typing a period on a blank line.
>
> Please reserve bridal suite for April 18th. Arriving
> San Francisco 5:12 a.m on the San Andreas Express.
>
> Warmest wishes,
>
> Eartha Quaker
> .

Send Options

Copies of your message can be sent to other users. They can be flagged as urgent, tagged for special handling, filed or, at your option, cancelled. You may also request that a message be mailed by the US Postal Service to persons outside the system. The following Copy options are available.

> **CC**   Courtesy copy
> **BCC** Blind copy
> **F**   Copy sent to folder
> **L**   Copy sent to list
> **C**   Copy sent to conference
> **B**   Copy sent to bulletin board

These alphabetic characters are entered at the "To:" prompt before the name of the addressee. Your command will be confirmed.

The Electric Mailbox

> To: CC Burks
> Copy of message addressed to Bob Burks

Messages can also be tagged in the address field for special handling using one or more of the following tags.

> **ANS** Answer
> **DEL** Delay sending message
> **NOF** No forwarding allowed
> **PAP** Send paper copy in U.S. Mail
> **REC** Return receipt
> **URG** Urgent

Tags are entered after the name of the individual to whom the message is directed and are preceded by a left parenthesis. More than one of the parameters may be entered at the same time. For example, to send an Urgent message, Return Receipt, you would make the following entry:

> To: Morris (URG REC
> Message addressed to Robert Morris

A return receipt is acknowledged in the sender's scan list by an entry of RCPT in the "Type" column of the scan list. The receiver of the message is notified of a receipt requested message by the following entry immediately below the posting.

> ****Receipt Request****

You can also stack multiple addresses and symbols on the same line, but each address must be separated by a semicolon.

> **To: Morris (URG REC; CC Daniels; BCC Fenley**

To add or change a tag, the name and tag must first be deleted by typing a minus symbol followed by the name and tag as it appeared when first entered. A confirmation is received of the deletion. Then type the new name and tag. A confirmation is received when the new entry is accepted.

> To: - BCC Smith
> Robert Smith removed from address list
> To: Roberts (URG
> Message addressed to: Margaret Roberts

Dialmail

Once a message is created, it can be reread, edited, mailed, cancelled, or filed. The following options can be used.

> **Read**   Review the just-created message.
> **Edit**   Edit the just-created letter. See the user's guide for a discussion of editing commands.
> **Edit Env**   Edit the address portion (envelope) of your message.
> **Send**   Send a created message. Type "SEND" at the Create menu prompt. Messages are assigned a unique message number that appears in the confirmation. This number can be used for tracking purposes should the message need to be traced.
> **Cancel**   Cancel a created message.

USING A MAILING LIST

Dialmail permits you to create permanent mailing lists that can be stored for use when writing to particular groups of people. To do this, type "CREATE LIST" at the main menu prompt. You will be asked for a title for the list. The title can contain up to 40 characters. Next, you will be prompted for the surnames you want the list to contain. The "Enter name:" prompt continues to appear until you press Return following a prompt without entering a name. Here is a sample file:

> Title: ABA Committee Members
> Please enter the names in the mailing list
> Enter name: Bryan
> George Bryan is now on the mailing list.
> Enter name: Woodard
> Nancy Woodard is now on the mailing list.
> Enter name: (Return)

To use the mail list, type "LIST" at the "To:" prompt followed by the name of the list. For example, to send mail to all members on a list called "ABA Committee Members," make the following entry at the "To:" prompt. You receive confirmation of the request.

> To: LIST ABA Committee Members
> Message addressed to: ABA Committee Members

SCANNING MAIL

The Scan command displays lists of messages, folders, and mailing lists, and permits a variety of actions to be taken on them. Type "SCAN" at the

"Your Command:" prompt. A list of scanning options is displayed. Inbox, Desk, Folder, and Lists are the scan commands most frequently used when working with email, and these are the ones we will discuss in this chapter.

Scan Inbox

A list of all messages currently in your mailbox is generated by typing "INBOX" and pressing Return at the command prompt. By reviewing the scan table, you can be selective in choosing the messages you want to read. This is an important consideration if you don't have the time to read all messages at the time you logon to the system. Messages remain in the Inbox for 30 days, but after that, they are automatically deleted from the system. Thus, it is wise to check your mailbox at least several times a month so that messages will not be lost.

The list is followed by a menu that displays the available options. You may choose to read, delete or print a message, or to scan for other messages.

Read   Read messages in the Inbox.
Delete   Erase a scanned message from the Inbox. Once deleted, the message is lost and cannot be retrieved.
Print   Printed messages offline at the DIALOG offices in Palo Alto, CA, and have them mailed to you via the US Postal Service.
Scan   Return to the Scan Menu to choose a Scan option. Re-enter "Scan" following the Inbox display.

Scan Desk

The Desk in Dialmail is used to work on current projects. Messages that have been read, but on which no action has been taken, are placed on the Desktop. This happens when "NEXT" is entered from one of the Inbox sub-menus. Dialmail automatically clears the desk each evening. Thus, any items remaining on it at the end of the online day should be handled prior to logging off since they cannot be retrieved the next day.

Type "DESK" at the Scan prompt to receive a list of messages currently on your desk, or type "SCAN DESK" from the main menu. You may choose to read or delete the message, Print it, or Scan for more messages. Notice that your options are the same as those offered when the Inbox command was entered. See the "Scan Inbox" section for a description of these options.

Dialmail

Scan Folder

Every Dialmail user has two folders already assigned by the system when they initially logon: Correspondence and Drafts. Other folders can be created. See "All About Folders" in the Dialmail Handbook. Type "FOLDER" at the Scan prompt to display a list of all current folders and the number of items in each one. The display is followed by the actions that can be taken on them.

    Scan   Scan the contents of a folder.
    Read   Read the contents of a folder.
    Delete   Delete the contents of a folder. As a precaution, you are asked to confirm your choice to delete it.

Scan Lists

Mailing lists that have been created in Dialmail can be reviewed and a variety of actions may be taken on them. Type "LISTS" at the Scan Menu prompt to display the names of your current mailing lists. The list is followed by the actions that can be taken on them. Let's examine the options you have available.

    Read   Review the names on a mailing list.
    Delete   Remove an entire list of names from your files.
    Add   Add names to a mailing list.
    Withdraw Remove a name from a mailing list.
    Send   Share mailing lists with other Dialmail users.
    Copy   Copy the contents of one list into another list.

READING MAIL

Mail is placed in your mailbox when it is received. Each time you logon to the system you are informed of the number of items in your Dialmail Inbox. To immediately begin reading your mail, type "READ" at the "Your Command:" prompt. You will receive a list of options from which you may choose to read unread messages in your Inbox or to read messages in your folder. Your Desk will not contain messages unless they were placed there on the same day as your current Dialmail session. This is because the Desktop is cleared automatically each evening.

Read Inbox

Use this command to read messages in the Inbox. You are prompted for the numbers of the messages you want to read. (The message number is

The Electric Mailbox

located in the far left column of the Scan list.) Numbers can be entered on a single line by placing a comma between each number; or as a range of numbers by inserting a hyphen between the range. To read every message in the list, type "ALL" at the prompt.

    4,5,8-10

Numbers can be entered in this way with the Read, Delete, or Print commands. Once the message is read, it can be reread, answered, deleted, or edited. Let's examine each of these options.

Read. To re-read the just-read message, type "READ" at the prompt and press Return. The message is displayed and is followed by a list of options you can take on the it.

Answer. This command provides a way for responding immediately to the sender of a just-read message. Type "ANSWER" at the command prompt and press Return. The system automatically enters the name of the person to whom the message is to be sent, since it is picked up from the received message. You may, of course, send the same message to another user by entering the name at the "To:" prompt. Otherwise, press Return to receive the Subject prompt. You may enter a new subject or simply press Return to automatically reference the subject of the received letter. Type the text, followed by a period on a line by itself and press Return. The Answer Menu option list is displayed. You may choose to review, edit, send, cancel, or file the message.

Delete. To delete a just-read message, type "DELETE" at the "Your command:" prompt. The message is erased and the deletion is confirmed.

File. The just-read message can be filed in an electronic folder by selecting this command from the menu. You will be prompted for the name of the file into which you want to place the message. After entry of the file name, the message is copied to the selected file and deleted from your Inbox. Confirmation of the action is displayed on the screen.

Edit. A just-read message can be edited the same as messages you compose yourself. You might want to edit a message if you plan to forward it to another individual and want to either add text or delete parts of the text before sending it. It is possible to edit not only the text of the message, but also the envelope (the person to whom the message is addressed). To edit the text, type "EDIT" at the prompt. To edit the envelope, type "EDIT ENV."

Dialmail

Next. Type "NEXT" at the prompt to display the next message in your mailbox. If no action has been taken on the just-read message, it is deleted from the Inbox and placed on your Desk. The just-read message can now be accessed by entering "DESK" from the Scan or Read menus.

Read Desk

Messages are placed on the Desk whenever they are read and no immediate action is taken on them. (See Scan Desk.) To read a message on your desk, type "DESK" at the Read Menu. You are informed of the number of messages on the desk, and then presented with them one at a time. Each message is followed by a menu of options from which you may choose to read, answer, delete, file, or edit the message. These are the same options described earlier in Read Inbox. Remember, messages remaining on your Desk overnight are automatically deleted from the system.

Read Folder

To read the contents of a folder, type "FOLDER" at the Read option prompt. You will be advised of the number and names of folders in your file and informed of the number of items in each one. At the folder prompt, type "READ." You are asked for the name of the folder you want to read, and then presented with the messages.

FILES AND FOLDERS

To create a file folder, type "CREATE FOLDER" from the Main menu, or at any time a File option is used. You will be prompted for the filename. If the name does not exist, you are asked if you want to begin one with the title you entered at the prompt. Your request is confirmed and you are returned to command level.

OTHER DIALMAIL SERVICES

Sending Paper Mail

Letters can be prepared online and then mailed via the US Postal Service to other Dialmail users or to persons outside the Dialmail system. The rate is 75 cents for the first page and five cents for each additional page. To use this option, type "CREATE" at the command prompt. At the "To:" prompt, type the full name of your correspondent since Dialmail will not necessarily have the name in its files. To let Dialmail know that you want a paper copy of the letter, type "(PAP" following the the name.

The Electric Mailbox

    To: Ms. Betty M. Roberts (PAP

Next, you receive the "Enter Address" prompt. You may enter up to four lines of no more than 50 characters per line. Dialmail uses a question mark ("?") to prompt for each address line. Press Return to signal the end of your address entry. A second "To:" prompt is generated. If you don't want to enter a second address, press Return to display the "Enter subject" prompt. The subject of the message may consist of up to 40 characters. End the subject by pressing Return. You may now enter your text, ending it by typing a period (".") on a line by itself and pressing Return. At this point, you may choose to read, edit, send, cancel, or file the letter.

LOGOFF PROCEDURES

To logoff the system, type "EXIT" at the Dialmail command level. A question mark will appear (the KI prompt) to which you respond by typing "LOGOFF."

CONCLUSION

Dialmail is a new electronic mail system that will probably undergo minor changes as the system is perfected. It is an easy system to learn, and if you are in need of a good bibliographic retrieval service with messaging capability, Knowledge Index and Dialmail are well worth your consideration.

MCI Mail

MCI Mail
MCI Digital Information Services
2000 M Street NW, Suite 300
Washington, DC 20036
202-293-4255
202-833-8484
800-624-2255

Most people know MCI as the long distance phone company that has been giving Ma Bell a run for her money for the past several years. Actually, discount long distance calling is just one of many communications services offered by MCI Communications Corporation.

MCI Mail was begun in 1983 and has since signed up nearly a quarter of a million subscribers. The relative success of the system is probably due in part to the versatility it offers; MCI was one of the first major systems to offer a paper mail option (other than Mailgrams or telegrams) and to give the common man access to the large worldwide telex network at a minimal cost. Now, MCI offers delivery of electronic and paper messages to anywhere in the world.

MCI Mail is a full-featured email system with many options. For example, if you need to send paper mail, either because the recipient is not a subscriber or you want the impact of an orange MCI Mail envelope, your message can be laser printed and delivered by first class mail or overnight courier. In some cities, a four-hour courier delivery is available.

You may choose to keep your letterhead and signature on file with MCI Mail so that they can be included in your paper mail automatically. MCI Mail even has special letterheads for holiday greetings, such as last-minute Christmas and Mother's Day letters.

The Electric Mailbox

HOW TO SUBSCRIBE

You may subscribe by calling MCI Mail at 800-MCI-2255 (or 202-833-8484 in the Washington, DC area.) You will receive a customer number, which is used for billing purposes, a user ID number, a user name, and a password. Your user name is normally your first initial and last name unless you specify otherwise. If you have a common name, such as John Smith, you may want to choose a unique user name, perhaps incorporating your company name, to differentiate you from other users with the same name.

RATES

There is an annual mailbox fee of $18.00 per subscriber, which is payable at the time of registration and billed on the anniversary date thereafter. There is no monthly fee or minimum for Basic Service. All charges are billed directly to the subscriber by MCI Mail.

Advanced Service is available for a $10.00 monthly fee. For the frequent user, the Advanced Service offers several conveniences, including additional commands, a brief command mode that avoids the use of menus, and additional online storage.

Charges for electronic Instant Letters sent to subscribers' mailboxes are: 45 cents for a letter up to 500 characters in length, $1.00 for a letter up to 7500 characters, and $1.00 for each additional 7500 characters.

For paper mail sent to US addresses, an MCI Letter delivered by first class mail is $2.00 for up to three pages. Overnight Letters delivered by courier are $8.00 for up to six pages. Four-hour Letters delivered by courier are $30.00 for up to six pages. For each of these options, each additional "MCI ounce" (7500 characters of text) is $1.00. Volume discounts are available. To send the same message to over 100 US addresses by MCI Letter or Overnight Letter, "broadcast pricing" applies. Depending on volume, an MCI Letter can be as low as $1.35, and an Overnight Letter can be as low as $5.80.

International delivery of paper mail is also available. The MCI International Letter costs $5.50 for a letter up to 7500 characters (about 6 pages) delivered anywhere in the world. For $12.00 to $30.00 (depending on the destination), the letter can be delivered by express courier to many countries, providing next day delivery to Europe and one to four day delivery to other parts of the world. (Delivery time varies with destination.)

MCI Mail

Domestic telex messages may be sent to MCII subscribers for 25 cents per minute, and for 43 cents per minute for subscribers to other carriers. Pricing for international telex messages varies by country. For a list of current rates, type "HELP TELEX PRICING" online.

There is no charge for connect time if you live in a city that has an MCI Mail local access number, although local phone company message unit charges may apply. (See the appendices for a list of numbers.) Subscribers in other locations may access MCI Mail through Tymnet; there is a five cent per minute communications surcharge for using Tymnet. An 800 WATS access number is also available. There is a 15 cent per minute surcharge for use of the 800 number.

To register a graphic, such as your letterhead or signature, there is a $20.00 annual fee per graphic. Your graphic is digitized and stored on MCI Mail's computer. It is then printed by the laser printer when you send paper documents via MCI Mail.

MCI Mail Alert, a service that telephones the recipient of your electronic message to let them know they have mail waiting, is available for a cost of $1.00 per alert.

Through a reciprocal agreement with Dow Jones News/Retrieval Service, subscribers may access Dow Jones indirectly through their MCI Mail account. Regular Dow Jones rates apply. Conversely, Dow Jones subscribers may access MCI Mail. (Dow Jones subscribers may use Tymnet or Telenet to access Dow Jones, which is a gateway to MCI Mail. Communications charges are included in the Dow Jones connect time charges.)

ACCESSING MCI MAIL

You may access MCI Mail through a local access number in approximately 50 major cities in the US, or via Tymnet from over 500 locations in the US, including Alaska, Hawaii and Puerto Rico. (See the appendices for a list of numbers.) An 800 number is also available for users in the US. Surcharges apply for Tymnet and 800 access. International users may contact MCI Mail at 202-833-8484 for a list of access numbers for over 50 countries. When travelling abroad, you should check with Customer Support before leaving the US to verify access numbers and communications protocols for the country you will be visiting.

The Electric Mailbox

CONNECTION PROCEDURES

If you are using a local MCI Mail access number or the 800 access number, dial the number and wait for the carrier tone. Press Return. You are ready to logon.

If using Tymnet, dial your local Tymnet number and wait for the carrier tone. Using 300 baud, the prompt "please enter your terminal identifier" will appear, and using 1200 or 2400 baud the prompt will be a series of garbled characters. Type the letter "A" but do not press Return. At the "please log in:" prompt, type "MCIMAIL" and press Return. You are ready to logon.

LOGON PROCEDURES

When you are connected to MCI Mail, you will receive a prompt telling you to "Please enter your user name." Type your user name and press Return. Remember that your user name is not the same as your ID number. It is normally your first initial and last name, or another user name you chose when you registered. For example, "JSmith" is a user name, whereas "123-4567" is an ID number.

At the "Password:" prompt, type your password, normally eight alphabetic characters, and press Return. You should see a message telling you that connection has been initiated.

USING MCI MAIL

The basic MCI Mail service is a menu-driven program. The optional Advanced Service allows you to enter direct commands without using menus. Since most beginners will start out using the Basic Service, the instructions in this tutorial will assume you are using the Basic Service menus. Additional commands available to Advanced Service subscribers are described in a separate section below.

Your MCI Mail mailbox has four areas: an Inbox, a Desk, a Draft area, and an Outbox. The Inbox is where unread incoming messages are stored. The Desk is where messages are placed once they have been read. The Draft area is where the letter you are currently editing is temporarily stored until you are ready to send it. The Outbox is where a copy of letters you have sent are placed for your reference.

Documents on your Desk or in your Draft area and Outbox are cleared away after they have stayed there for 24 hours, so if you need to keep a

MCI Mail

copy of these documents, be sure to download a copy to your disk or printer. Your Inbox stores unread messages indefinitely.

MCI Mail will work with a wide variety of terminal types. To adjust the line length and number of screen lines of displayed messages, the time zone used in the header of your messages, and other options related to your account, type "ACCOUNT" at the Main Menu command prompt. Because of the way MCI Mail formats text for paper mail, a line length of 72 characters is recommended, but you may choose the setting that best suits your terminal.

Obtaining Help

MCI Mail offers extensive online instructions. To see a list of available commands, type "HELP" at most any prompt. For help on a particular command, type "HELP" followed by a space and the command. For a list of available help files, type "HELP INDEX" at the Main Menu prompt (Basic Service) or the Command prompt (Advanced Service). You may call the Customer Support department between the hours of 9 a.m. and 8 p.m. Eastern Time, Monday through Friday, at 800-424-6677. Or, you may send a toll-free MCI Mail Instant Letter to user name "MCI HELP" with questions.

Special Control Keys

The following keys may be used for editing and flow control on MCI Mail.

CTRL H	Backspace. Delete a charater.
CTRL W	Word erase. (Does not work on Tymnet.)
CTRL R	Redisplay a line.
CTRL X	Line erase. (Does not work on Tymnet.)
CTRL S	Freeze display.
CTRL Q	Resume display.
CTRL L	Page break. Used when editing text for paper mail.

SENDING AN INSTANT LETTER

Sending a letter on MCI Mail is very simple. When you see the Main Menu, you will be prompted for "Your command." The various options available are self-expanantory. To send a letter, you must first tell MCI Mail that you wish to create a message. Type "CREATE" and press Return. At the "To:" prompt, type the user name or user ID number of the MCI Mail subscriber to whom you wish to send the letter.

The Electric Mailbox

```
SCAN        for a summary of your mail
READ        to read messages one by one
PRINT       to display messages nonstop
CREATE      to write an MCI letter
HELP        for assistance
EXIT        to leave MCI Mail
```

Your command: CREATE

To: 123-4567
  Norman Bates Bates Motel Mesa CA

If you enter the user ID number, the name of the user will appear below your entry. If, however, you type the user name and there is more than one subscriber using that name, a list of choices will appear. You must specify which person you want.

To: NBates

There is more than one:

No.	MCI ID	Name	Organization	Location
0	NOT LISTED BELOW. DELETE.			
1	NOT LISTED BELOW. ENTER AN ADDRESS.			
2	123-4567	Norman Bates	Bates Motel	Mesa CA
3	234-5678	Nancy Bates	ABC Corp.	Reno NV

Please enter the number:2

If the correct person is listed, select the correct number from the left column. If you press "1" you are prompted to enter a street address for paper mail delivery. For now, we'll learn how to send an electronic Instant Letter.

After you enter the user name or ID, you will receive another "To:" prompt. You may continue to enter additional addresses. When you are finished, press Return at a "To:" prompt. You will then see the "CC:" prompt. If you wish to send a courtesy copy to someone, enter their user name or ID at this prompt. Otherwise, press Return. To cancel a message, type a slash at a "To:" or "CC:" prompt.

You should now see the "Subject:" prompt. Type a subject line and press Return.

MCI Mail

    To: (Return)
    CC: (Return)
    Subject: Accommodations

Now you will be prompted to enter the text of your message. Lines should be no longer than 80 characters, and you should end each line by pressing Return. When you are finished, type a slash ("/") on a line by itself to end.

    Text: (Type / on a line by itself to end)

    Dear Mr. Bates:

    I will be arriving the evening of Friday, December 11th.
    Please reserve a single room with private shower. I will
    only be staying for one night, and I will pay cash.

    Thank you,

    Marion Crane
    /

Next you will see the Send Menu offering you choices of what to do with the draft of your letter. You may edit it, read it, or send it.

READ	to review your letter
EDIT	to edit your letter
SEND	US Mail for paper or instant electronic delivery
SEND ONITE	Overnight courier for paper; Priority electronic delivery
SEND 4HOUR	Four-hour courier for paper; Priority electronic delivery
HELP	for assistance

    Command (Menu or Exit): SEND

To send the message, merely type "SEND" at the prompt, and press Return. A message will be displayed verifying that your letter was sent.

    Your message was posted: Thur Dec 10 1986 2:59 pm PDT
    There is a copy in your Outbox.

The Electric Mailbox

EDITING A LETTER

If you need to edit a letter prior to sending it, type "EDIT" at the menu you see after creating your letter. To edit a draft from the Main Menu prompt, type "EDIT DRAFT" and press Return.

The Edit Menu allows you to make changes in the text of the letter or the "envelope," which contains the address and subject fields.

>Your command: EDIT
>
>You may enter:
>
>| | |
>|---|---|
>| ENVELOPE | to edit the TO, CC or SUBJECT fields |
>| TEXT | to edit the TEXT field |
>| LEAVE | when you are finished editing |
>| HELP | for assistance |
>
>Your command: ENVELOPE

If you want to add or delete addresses, edit the envelope field. Type "ENVELOPE" at the Edit Menu. You will receive the "To:" prompt, at which you may type an additional address. To delete an address type a minus sign (-) followed by the name you wish to have removed. If you press Return at the "To:" prompt, you are taken to the "CC:" prompt, where you may add or delete addresses for courtesy copies in the same manner. At the "Subject:" prompt, you may either enter a new subject line or press Return to leave the subject unchanged.

To edit the text of your message, type "TEXT" at the Edit Menu. You are given the choice of reading your message, adding or deleting lines of text, or changing text. Since the editor uses line numbers, you may need to read all or part of your letter before changing text. To do this, type "READ" followed by a range of line numbers, or "READ ALL" to see the entire letter.

>Your command: TEXT
>
>You may enter:
>READ, ADD, DELETE, CHANGE, or STOP
>
>Your Edit command: READ 1-5

MCI Mail

To add lines to the end of the letter, simply type "ADD" at the edit command prompt. To add text between lines, type "ADD" followed by the new line number. (You may use decimals. For example, "ADD 3.1" will add a line of text after line three.) You will be prompted to enter text. When you are finished, type a slash ("/") on a line by itself and press Return.

To change text, type "CHANGE" followed by a line number at the edit command prompt. You will be asked for the "old text" you wish to change and the new text you wish to replace it with. If you want to delete lines, type "DELETE" followed by a line number or a range of line numbers at the edit command prompt.

To format text for paper mail, use the "READ PAPER" and "EDIT PAPER" commands. These commands allow you to see the number of lines that will be printed on each page and the width of each line of text.

When you have finished editing your letter, type "STOP" at the Edit command prompt. You will be returned to the Edit Manu. Type "LEAVE" to go to the Send Menu. To send your letter, type "SEND" and press Return.

Sending Options

In addition to the electronic Instant Letter, MCI Mail gives you the option of specifying priority delivery and/or paper mail delivery. To send an electronic Instant Letter to an MCI Mail subscriber with priority handling, select the "SEND ONITE" or "SEND 4HOUR" option from the Send Menu. The message will appear in the recipient's mailbox with a "**PRIORITY**" flag to let them know it is an urgent letter.

When you enter a name at the "To:" or "CC:" prompt that is not a registered MCI Mail user name, you are asked for an address. You may enter a street address for paper mail delivery, a telex address, or the address of another electronic mail service that will receive messages from MCI Mail.

```
0 - DELETE
1 - Enter a PAPER address
2 - Enter a TELEX address
3 - Enter an EMS address
```

Please enter the number:1

To send paper mail, select "1" and answer the prompts for the address. See the sections below for information on sending telex messages and messages to other electronic mail services.

The Electric Mailbox

If you wish to send paper mail to an MCI Mail subscriber, type "(PAPER)" after the user name or ID number.

> To: 123-4567 (PAPER)

Unless you specify priority delivery, paper mail letters will be laser printed by MCI Mail and sent in an orange envelope via first class mail through the US Postal Service.

For overnight or four-hour courier delivery, select the "SEND ONITE" or "SEND 4HOUR" options from the send menu. These services are available only in certain areas. For a list of zip code areas served, type "HELP" followed by a space and the state to which your letter is to be delivered.

> Your command: HELP NEW JERSEY

Remember that couriers cannot deliver to post office boxes and must have a complete street address. It is also a good idea to include a phone number in the second address line of your letter.

If your letter is addressed to several different people and you do not want priority delivery for all addresses, you may specify the sending option after the user name.

> To: JDean (ONITE)

If you post Overnight Letters by 11 p.m. Eastern Time, they will be delivered by noon the next day. Four-hour Letters posted by 6 p.m. the recipient's local time will be delivered within four hours.

Receipt Requested. To request a receipt verifying delivery of your message, type "(RECEIPT)" after the user name at the "To:" or "CC:" prompt.

> To: Norma Desmond (RECEIPT)

For Instant Letters, a notice will appear in your Inbox indicating the date and time your letter was read. For paper mail, the receipt indicates the date and time that your letter was prepared for delivery. Four-hour Letters automatically provide you with confirmation of the date and time your letter was delivered and the name of the person who signed for the delivery.

Telephone Alert. If you are sending an urgent message to an MCI Mail subscriber in the US, and you want to be sure the recipient checks for mail, you can have an MCI operator telephone the recipient to remind

MCI Mail

them to check their mailbox. The operator will attempt the call three times at half-hour intervals. No calls will be made after 7 p.m. Eastern Time. You will receive a notice in your MCI mailbox telling you if the recipient was reached or not. You will receive notification of delivery when the recipient reads the message.

To request a telephone alert, type "(INSTANT, ALERT: phone number @ time)" after the user name at the "To:" prompt. Notice that you may specify the time to call.

> To: Bob Woodward (INSTANT, ALERT:201-555-4321 @ 4:00 p.m.)

There is a $1.00 charge per message for this service.

Charge Codes. If you need detailed information about your electronic mail communications for cost accounting, tracking by department or project, or client billing, the charge code option will make your life easier. At the "To:" or "CC:" prompt, type "(CHARGE: code)" after the user name. The code can be any any alphanumeric code from three to 20 characters in length. Do not include spaces, hyphens or symbols, except the dollar sign ($).

> To: Herbert Hoover (CHARGE:DAMPROJECT)

When you receive your monthly invoice, correspondence will be categorized by charge code. If several user IDs are billed on one invoice, you may request that charge codes be permanently assigned to each ID. That way, each ID's correspondence will be grouped together on the invoice.

Signatures and Graphics. You may register a graphic, such as a letterhead, and/or a signature to be included in your paper mail letters. Basic Service subscribers may register only one graphic and one signature, but Advanced Service users may register up to 15 of each. To include your signature in the text of your letter, type "/SIGNED/" on a new line where you want the signature to appear, and press Return. Your registered letterhead will be used on all of your paper mail. However, Advanced Service users may specify the letterhead to be used.

USING A MAILING LIST

If you regularly send messages to the same groups of people, you may want to use the List option. To create a new mailing list, type "CREATE LIST" at the Main Menu command prompt. You are prompted for the name you wish to give the list. The name must be alphanumeric, from three to

The Electric Mailbox

20 characters in length, and contain no spaces. At the "To:" prompt, you may begin entering addresses. Included handling options, such as Instant, Receipt, Onite, etc., in parenthesis after each user name. To review what you have typed so far, you may type "*R" and press Return at the "To:" prompt. Press Return at the "To:" prompt to end the list.

> Your command: CREATE LIST
> LIST name: DIRECTORS
>
> To: C.B. DeMille (INSTANT)
> To: E. VonStroheim (RECEIPT)
> To: (Return)
>
> Your address LIST DIRECTORS has been registered.
> There are 2 members in this list.

To read your list, type "READ" at the Main Menu command prompt, then type "LIST" at the next prompt. To edit the list, type "EDIT" then type "LIST." You may also change the name of the list when you edit it. To delete a list, type "SCAN" at the Main Menu, then type "LIST," then "DELETE." (Advanced Service subscribers may combine commands on one line.)

To send mail to the addresses on your list, just type the name of the list at the "To:" prompt of your letter.

UPLOADING TEXT TO MCI MAIL

Unless you normally send short notes or replies, you will probably find it most convenient to compose your letters offline using your word processor. You should save text as an ASCII text file, with no special control codes or formatting codes. A right margin of 72 characters or less is best because of the way MCI Mail formats letters for paper mail. Include a carriage return at the end of each line (no line feed). When you receive the "Text:" prompt, take whatever steps are required by your communications software to begin transmitting the file. Be sure that Xon/Xoff is engaged for both transmitting and receiving. Don't forget to type a slash on a new line to indicate the end of your letter. To be sure the file transmitted properly, you may wish to read your draft before sending it.

READING MAIL

Probably the first thing you will want to do when you logon to MCI Mail is to scan your mailbox for any new messages. At the Main Menu, type

MCI Mail

"SCAN" and press Return. When you receive another prompt, type "INBOX" and press Return. If you have any unread messages, a scan list will appear.

No	Posted	From	Subject	Size
1	Dec 14 9:00	Rose Woods	Recordings	240
2	Dec 14 9:10	Don Lockwood	My contract	560

To read your messages, at the Command prompt, type "READ" and press Return. You will be prompted for the scan numbers for the messages you wish to read. Type a number, a range of numbers separated by a hyphen, or a list of numbers separated by commas, and press Return.

After you read a message, you may type "NEXT" and press Return to read the next message in your Inbox, or simply press Return if you specified a list or range of message numbers.

If you wish to reply to a letter, at the Command prompt that appears after you read the letter, type "ANSWER" and press Return. If you want all the original recipients to receive a copy of your reply, type "ANSWER ALL" instead. In either case, the envelope is created for you automatically. You are prompted to enter the text of your reply. When you are finished, type a slash ("/") on a new line and press Return.

If you wish to have your mail displayed continuously without "Press Return for More" prompts at the end of each screenful of text, use the Print command instead of the Read command. This is especially useful if you are downloading text to be printed out.

ADVANCED SERVICE COMMANDS

Advanced Service provides several options not available to Basic Service subscribers. First of all, draft, desk and outbox messages may remain in your mailbox for up to five days instead of 24 hours. Advanced Service users have a Pending area that can hold several drafts at one time. And it saves time using the command mode provided by Advanced Service instead of waiting for menus to scroll up the screen for each subcommand, as is required by Basic Service.

Instead of menus, Advanced Service users simply receive a Command prompt. Commands may be combined on one line. For example, you may type "READ INBOX" to read your new messages, "SCAN PENDING" to scan a list of your draft messages, or "EDIT TEXT" to edit the text of your letter.

The Electric Mailbox

Advanced Service provides a "Handling" prompt when you end the text portion of a letter. Various sending options, including Charge, specification of signature and letterhead, Receipt, and others may be specified at this point. Multiple options can be included on one line if they are separated by commas. You are asked if you wish to send the letter. If you answer "N," the draft is placed in your pending area. To add handling options, type "EDIT ENVELOPE" at the Command prompt until you get to the "Handling" prompt.

A major option available only to Advanced Service users is the Forward command. To forward a message you have just read, type "FORWARD" at the first Command prompt. You will receive a "To:" prompt, just as if you were creating a letter. When prompted to enter text, you may type remarks to be used as a cover note to the forwarded message. Type a slash on a new line to stop entering text.

You can also forward messages from the scan list. At a Command prompt, type "SCAN ALL" to see a list of messages in your mailbox. Then type "FORWARD" followed by a space and the numbers of the messages you wish to forward.

Another powerful Advanced Service option is the Include command. You may include the text of another message in your mailbox in a message you are composing. Type "/INCLUDE/ message indentification" at the point where you want the text included. You may identify the message by scan number or a combination of date, name and subject.

```
     Emily: Here is a note I received from the insurance people
     in Topeka about our policy on the farm house.
     /INCLUDE/DESK FROM GULCH
     Text from the following message will be copied into this draft:
          Posted          From              Subject        Size
       May 18 1986      Gulch Ins.        Policy lapse     312
     Please check into this for me, will you? Love, Henry.
     /
```

SHORTCUTS

Most MCI Mail commands can be abbreviated to the first two letters of the command, or enough letters to make the command unique. For example, you may type "CR" for "CREATE," "RE" for "READ," or "IN" for "INBOX."

As mentioned above, using the command mode available to Advanced Service users is a great time-saver if you access MCI Mail on a daily

MCI Mail

basis. By combining commands on one line and avoiding repetitive menus, you can complete your online session with less effort.

It usually saves time to use a recipient's ID number instead of their user name if there is more than one subscriber with the same user name. If you are not sure of the exact spelling of a user name, or you wish to verify the ID number, use the Find command. Type "FIND" at the Main Menu prompt. (Advanced Service users type "FIND NAME" at the Command prompt.) When prompted for a name, you may enter a name in one of seven different ways. Use one of the following formats:

Full name	Dorothy C. Gale
Last name	Gale
Initial and last name	D Gale
Name and organization	Gale/Ruby's Shoes
Name and location	Gale/Topeka
Name, location, organization	Gale/Topeka/Ruby's
User name	DGale

A list of subscribers, along with their ID numbers, that match your search will be displayed. If you find the one you need, jot down the ID number for future use.

OTHER SERVICES

Sending Telex Messages

MCI Mail subscribers may send and receive telex messages to and from most any telex system worldwide. To send a telex message, type "(TELEX)" after the user name at the "To:" prompt. You will be prompted for the necessary information. For domestic telex delivery, press Return at the "Country:" prompt and select the appropriate telex system, telex number, and answerback. The answerback is optional.

```
Country: (Return)

No. CODE   COUNTRY NAME
0          NOT LISTED BELOW. DELETE.
1   -      MCI/WUI (OR OTHER IRC TELEX NETWORK)
2   0      WESTERN UNION TELEX I NETWORK
3   5      WESTERN UNION TELEX II (TWX) NETWORK

Please enter the number: 2
Destination is WESTERN UNION TELEX I NETWORK (0).
```

The Electric Mailbox

>Telex No: 0-234567
>Answerback: ABRACAD

For overseas telex, at the "Country:" prompt, type the name of the country. The country code will be inserted automatically by the system. Type the remainder of the telex number at the prompt, then type the answerback when requested. You will be asked to verify the telex address you have typed.

>Country: Australia
>Destination is AUSTRALIA (790)
>
>Telex No: 790-

You will automatically receive a receipt in your Inbox telling you that your telex message was delivered or that it was cancelled due to non-deliverability.

You may prepare text for telex messages in the same way you prepare other MCI Mail letters, except that there may not be more than 69 characters per line and each message must not exceed 12,000 characters.

Receiving Telex Messages

Telex subscribers can send messages to your MCI mailbox. Your MCI Mail telex number is determined by placing the numbers "650" in front of your ID number. Your answerback is this number followed by a space and "MCI." For example, if your MCI Mail ID number is 123-4567, your telex number would be "6501234567" and your answerback would be "6501234567 MCI" for incoming telex correspondence. Inside the US, correspondents should enter "101" (except MCII/WUI subscribers), then at the "WUI GA" prompt, they should type your MCI Mail telex number. Outside the US, correspondents must use the MCI/WUI Access Code as a prefix to your telex number. Type "HELP TELEX INBOUND" at a command prompt for a list of Access Codes by country of origin.

CompuServe Messages

MCI Mail subscribers may exchange messages with CompuServe EasyPlex and InfoPlex subscribers. Refer to the chapters on EasyPlex and InfoPlex for instructions on sending messages from CompuServe to MCI Mail.

MCI Mail

To send a message from your MCI Mail account to a CompuServe subscriber, at the "To:" prompt, type the correspondent's name followed by "(EMS)" and press Return. You will be prompted for the name of the electronic mail system.

    To: Melanie Daniels (EMS)

Type "CompuServe" and press Return. At the "MBX" prompt, type the recipient's CompuServe ID number or user name. At the next "MBX" prompt, press Return to end address entry. Each CompuServe recipient must be entered at a separate "To:" or "CC:" prompt.

LOGOFF PROCEDURE

Type "EXIT" at any Command prompt to logoff MCI Mail.

CONCLUSION

MCI Mail offers a wide variety of features suitable for anyone needing both electronic and paper mail delivery, including large companies with sophisticated communications needs. Yet it is accessable to anyone, including individuals and small businesses with low volume requirements. By offering two-way access to the telex systems and to other email services such as CompuServe, MCI Mail opens up a world of communications options at a reasonable cost. The system is simple enough for the novice and flexible enough for the power user. If MCI continues its efforts to interconnect with other electronic mail services, MCI Mail should be a major contender in the email market for years to come.

EasyLink

EasyLink
The Western Union Telegraph Company
One Lake Street
Upper Saddle River, NJ 07458
800-336-3797

EasyLink, Western Union's high-speed store and forward electronic mail service, was introduced in 1982 on an experimental basis and has grown rapidly since then. It now boasts over 140,000 users, sending upward to 2 million messages daily. Messages are sent outside the EasyLink network by a variety of Western Union electronic mail products. Domestically (United States, Canada, Mexico, Puerto Rico) messages can be sent by Telegram, Mailgram, and Computer Letter. Cablegrams and Overseas Priority Letters are available for foreign delivery.

EasyLink also offers a number of business and consumer databases through its FYI service. You have access to Official Airline Guides (OAG), UPI news bulletins, and the NYSE, ASE, and NASDAQ stock exchanges. InFact, an electronic information service, was added in 1985. It combines the resources of over 630 databases, including BRS, Dialog, and NewsNet. A new service provided by EasyLink is "Online Conversations." It allows you to communicate directly with Telex I, Telex II, and WorldWide Telex terminals. Instead of sending messages to your correspondent, you are connected directly to his terminal and type your message while online, or upload it from disk or tape. Your correspondent can reply directly to you using EasyLink's switchboard.

The United Kingdom has an EasyLink mail system similar to Western Union's system that is owned and operated by Cable and Wireless, Ltd. Messages can be sent to and received from United Kingdom subscribers

EasyLink

by Western Union EasyLink users. All features available to EasyLink subscribers can be used when sending messages to the United Kingdom.

EasyLink publishes its own software called "Instant Mail Manager," but it is not required to use the system. Currently, the software is available only for IBM PCs and compatibles. Versions for other systems are planned.

HOW TO SUBSCRIBE

Make application by calling 800-336-3797, ext. 508, and asking to speak to an EasyLink representative. As a new subscriber to EasyLink, you receive a welcome kit that contains your EasyLink ID, user name, password, mailbox number, Telex number, and terminal type code. You must have this information to logon to the system and use its services.

RATES

EasyLink operates 24 hours a day, 7 days a week. It offers two subscription fee plans: annual and monthly. For an annual subscription fee of $25.00 you are permitted to send any number of messages with no monthly minimum usage imposed. On the monthly plan, a $25.00 monthly minimum usage is required. The monthly usage minimum is not effective until after your first 30 days of service. A registration fee is not charged.

All charges are calculated in six second or full minute increments depending on the type of message sent. The subscriber is billed for any portion of an increment as if it were a full increment. Extra charges are made for other EasyLink features such as additional mailboxes and notification of delivery of a message. EasyLink supports baud rates 110, 300, and 1200.

Rates for EasyLink and Telex follow. They are in effect 24 hours a day. Usage rates for Mailgrams, Cablegrams, and other services offered by Western Union are discussed in the sections dealing with them.

From metro or local access numbers, Mailbox messages are 35 cents using 300 baud, and 50 cents using 110 and 1200 baud. There is an additional 20 cent fee per Mailbox address. Telex messages are 50 cents for all baud rates, and for Mexico, Canada, and other non-Western Union carriers, the rate is 70 cents for all baud rates.

For WATS callers, Mailbox messages are 60 cents for 300 baud, and 75 cents for 110 and 1200 baud. There is an additional 20 cent fee per Mailbox address. Telex messages are 75 cents for all baud rates, and for Mexico, Canada, and other non-Western Union carriers, the rate is 95 cents for all baud rates.

Telex rates are 50 cents for Mailbox messages, plus an added fee of 20 cents per address; 50 cents for Western Union Telex, and 70 cents from Mexico, Canada, and other non-Western Union carriers.

EasyLink Mailbox messages are based on input time. All other rates are based on output time which is approximately 400 cpm.

WATS users have additional charges as follows: connect charge, a 25 cent flat fee each time a connection is made; mailbox retrieval, 25 cents a minute billed in 6 second increments; and mailbox hold retrieval, 25 cents a minute billed in 6 second increments.

The cost for sending messages to the United Kingdom is 25 cents per 1000 characters plus a 40 cent message charge. Off-peak hour discounts do not apply.

Volume discounts are available as follows: 10% discount for monthly usage in excess of $1,000; up to $100.00 discount for monthly usage in excess of $500.00; and 40% discount for off-peak use of the system. The latter discount is available for EasyLink Mailbox and Telex users weekdays between midnight and 7 a.m. Eastern Standard Time, and applies all day Saturday, Sunday, and national holidays.

ACCESSING EASYLINK

EasyLink is accessed by an EasyLink "connect" number, available in metropolitan areas, or a local access number. See the appendices of your EasyLink User's Guide for a list of numbers.

CONNECTION PROCEDURES

When a connect number is used to access EasyLink, you will know you are connected when you see the "ID?" prompt. You can then proceed to follow logon procedures in the next section.

When EasyLink is accessed using a local access number, you will be prompted first for a terminal identifier. Your response depends on the baud rate setting. Type an "A" for 300 or 1200 baud or a "D" for 110 baud, and press Return. Do not leave a space between the prompt and your entry. Sometimes when using 110 or 1200 baud, the prompt appears garbled, but do not be concerned with this. The system will then prompt

EasyLink

with "Please Log In:." Press CTRL P and type "ESL" and press Return. If your terminal is set at half duplex, your ESL entry is displayed as EESSLL. This is normal and in no way invalidates your entry. You are informed when the call is connected and then prompted for your ID. You will see the message "Western Union Call Connected."

LOGON PROCEDURES

When you are connected to EasyLink, at the "ID?" prompt, type your customer ID. Your ID consists of a terminal type, EasyLink ID, username, and password. It is formatted like this:

    ID? Terminaltype (Space) EasyLinkID (Space) Username (Period) Password (Return)

Note that a space is not entered following the question mark or between the username and password. Include spaces only where indicated. After Return is pressed, the session number, date, and time will be displayed. The session number changes each time a command is fulfilled and a new PTS response is presented. Your ID entry looks like this:

    ID?08 XYZ13691 SSMYTH.BLT (Return)

If you selected full duplex, your ID may not be seen on the screen. In these situations, be careful when making your entries since you will not be able to see your mistakes. EasyLink allows you to make three attempts at entering your ID before automatically disconnecting you from the system. If this happens, simply redial and login again.

USING EASYLINK

EasyLink is basically a command driven program. All commands are entered on the line following the PTS response, and each command is preceded by a slash ("/"). PTS is the system's way of telling you that it is waiting for a command. EasyLink provides a Prompt mode that is helpful to the new or occasional user. To enter the Prompt mode, type "/PROMPT" and press Return at the PTS response. You will receive a menu of options from which to choose the function you want to perform. For example, if you type "1" (Send) at the "Enter A Number" prompt, you are prompted for each required entry for sending a message. EasyLink's prompt feature makes sending or reading any type of message easy. Just pick a number and follow the clear and explicit instructions. Once you feel comfortable with EasyLink, the prompt mode will become bothersome.

The Electric Mailbox

You can then return to the command mode, confident that help is always available if you need it.

Obtaining Help

Help is available online by typing "/HELP" at any PTS response. A list of all available help messages is displayed from which you can choose the item with which you need assistance. There is no charge for this service. Additionally, customer service is available 24 hours a day by calling 800-982-2737. Of course, you can always consult the user manual for assistance.

Special Control Keys

The following keys may be used to make corrections on the line you are currently typing.

    CTRL H   To delete a character.
    CTRL X   To delete a line.
    BREAK   To delete a previous entry. Cancels instructions, or discards a message.

If you do not have a Break key, type "+EEEE" following a PTS response and press Return. Telex I users should type "EEEE" and press Return.

SENDING A LETTER

EasyLink messages are called "Mailbox Messages." To send a mailbox message, on the line following the PTS response, type the mailbox number of your correspondent, a "+," and press Return. The "+" signals Easylink that the number is complete and that you are ready to begin the message.

    6246810+

The GA ("go ahead") prompt is displayed. This prompt tells you that the message can be entered. Type the text and press Return when it is complete.

A mailbox message cannot exceed 200,000 characters, including address and text. If a line exceeds 80 characters, EasyLink word wraps the text. At this point you have several options. You can send the message and disconnect from EasyLink. To do this, type "MMMM" and press Return. You can send the message and stay connected. Just type "LLLL" and press

EasyLink

Return. Or you can delete the message without sending it. To delete, press the Break key, or type "+EEEE" and press Return if you do not have a Break key.

>   987656X   21OCT86 11:45 EST
>
>   PTS
>   62938746+
>
>   GA
>   LB:
>
>   I am so sorry that I won't be able to make it to
>   the premier of "Camille" tonight, but I have developed a
>   slight cough, and I want to be alone.
>
>   Greta
>   LLLL
>
>   ACCEPTED 423186
>   EASYLINK
>
>   987656X   21OCT86 11:48 EST
>   PTS

When a message is accepted for transmittal, the system responds with "Accepted" followed by a message number and the word EasyLink as shown above. If the message is rejected, the system displays "Rejected" followed by the message number and the reason for rejection.

You may find it practical at times to forward received or created messages to another correspondent. Messages are only available for forwarding for five days after they were sent. To forward a message, you must have the message number that was assigned by EasyLink at the time it was sent. On originated messages, the number appears just below the LLLL or MMMM entry made at the end of the message.

>   LLLL
>   ACCEPTED 987654A991

And on received messages, the message number appears at the top of the message and is preceded by the letters "MBX."

>   EASYLINK MBX 923452A991

The Electric Mailbox

To forward a message, type "/FWD," a space, the message number, and press Return. On the next line, type the address, a "+," and press Return. The GA response is then displayed. Since you are forwarding text and not adding to it, type LLLL or MMMM below the GA prompt to send the message.

To add text to a forwarded message, make address entries as above, type the text following the GA prompt, and press Return. Then type the appropriate sending option, LLLL or MMMM, to send the message. The added text will appear at the beginning of the forwarded message.

Send Options

Now, let's take a look at some of the mailing options available when using EasyLink. Many of these options can be used when sending other types of messages such as telex messages and Mailgrams. You will be informed which services permit their use.

EasyLink provides two special mailing options: Notification of Delivery and Priority Delivery. These options can be used only with Mailbox, Telex, and WorldWide Telex numbers, or with abbreviated addresses that represent mailbox or Telex numbers. The options may be combined in one message.

Notification of Delivery (/NTF). Notification of Delivery serves the same purpose as a post office "return receipt." When a message is sent using this option, you are notified by a message in your mailbox confirming the delivery. To request Notification of Delivery, type "/NTF" and press Return. Then enter the address and message in the usual way.

    /NTF (Return)
    62843212+ (Return)

There is a charge of 25 cents for mailbox and Telex notifications, but no charge for WorldWide Telex. An additional fee of $1.00 is made for overnight notifications.

Priority Delivery (/PRI). Priority Delivery provides an easy way to send an urgent message. EasyLink makes every effort to deliver priority messages within 30 minutes of your entry of the message. If delivery cannot be made in this time frame, you are sent a cancellation notice. You may resend the message or you may choose to contact your correspondent by some other means. To request Priority Delivery, type "/PRI" and press Return. Then enter the address and message in the usual way.

EasyLink

    /PRI (Return)
    62993388+ (Return)

A 20% surcharge is added to the cost of the message. To send a message with both Notification of Delivery and Priority Delivery, type "/NTF PRI" on the line below the PTS response.

Attention Line (/ATTN). Attention lines are used to send messages to a particular person or to serve as a reference to the contents of the message. With the exception of Mailgrams, the attention line is entered as the last line in the address. With Mailgrams, it is entered on the line immediately preceding the message.

To send a message with an attention line, type "/ATTN," a space, the name of the individual or the reference information, a "+" and press Return. The attention line can contain up to 60 characters of text.

    /ATTN Sarah Roberts+ (Return)
    /ATTN XYX Conference Rescheduled+ (Return)

Attention lines may be used with Mailbox messages, Telex, WorldWide Telex, and Mailgram messages. They can also be stored with abbreviated addresses and in RediLists.

EasyLink provides several address options that assist in managing your mail more effectively and efficiently. Let's briefly review each of these. For more information, consult the "EasyLink Features" chapter of your user manual.

Multiple Addresses. The Multiple Address option permits you to send the same message to many individuals using a variety of message types. For example, a single message can be sent in one session by Cablegram, Mailbox, Mailgram, Telegram, and WorldWide Telex to many people. RediLists and abbreviated addresses can also be used with this option.

Alternate Addresses (/ALT). Alternate addresses are helpful in the event EasyLink has a problem delivering your message to the address of choice. An alternate address can be used with Mailbox messages, Telex, and WorldWide Telex, and can be stored with abbreviated addresses or in a RediList. Priority Delivery and Notification of Delivery are permitted.

The Electric Mailbox

Abbreviated Addresses

Entering commands and addresses every time you send a message can be bothersome and time consuming, particularly if your address list is long. EasyLink provides a simple way to avoid this laborious task. A two digit code can be used to represent an individual prestored address. It is possible to format up to 100 addresses in this way.

Abbreviated addresses may be used for all EasyLink messages except Computer Letters. They may be used for a single address, multiple addresses or batch input. Individual attention lines are permitted for those types of messages that accept them.

USING A MAILING LIST

EasyLink's RediList provides a way to store up to 250 addresses in a file for use with frequently used correspondents. The lists may be used with a single address, multiple addresses, or batch input. You can have as many lists of up to 250 addresses as you want, but each must be under a different RediList code. All message types, except Computer Letters, can be used in RediLists, and attention lines are permitted if allowed by the message type.

Each list is assigned a code name that can contain from 1 to 8 alphanumeric characters. The first character in the code must be a letter. "A123," "ABTCOMM," "GOODGUYS," and "Z1Y3" are all acceptable RediList names.

Request RediList service by calling EasyLink's Customer Care Center at 800-WU-CARES or sending a message to /SVC. They will help set up your list.

To send a message using a RediList, on the line below the PTS response, type the RediList code, a "+," and press Return. At the GA prompt, type your message in the usual way.

    PREUROPE+ (Return)

UPLOADING TEXT TO EASYLINK

Messages can be prepared online; however, you may find it more cost-effective to create them ahead of time, particularly if they are long, and then use the /Batch command to upload them to EasyLink.

EasyLink

Files must be stored as standard ASCII text files. Do not use boldface, underlining, or tab settings. The underline key may only be used to indicate a line, and use the space bar to set up columns. A batch input may contain up to 2 million characters. In all instances, you must adhere to the line width and message length for the type of message being sent.

You have two options when uploading to EasyLink: you can connect to EasyLink, enter the addresses and then use the /Batch command to send your prepared message; or, you can create addresses and messages offline, save them to disk or in a memory buffer, then connect to EasyLink and upload both addresses and messages.

To type addresses online and send messages using batch input, logon to EasyLink in the usual way. After the PTS response, type the address(es) to which the prepared message is to be sent, and type "+" following the last address entry. Following the GA prompt, type "/BATCH" and press Return. Now you are ready to enter the commands or keystrokes required by your software to upload the message. After the message is uploaded, type the appropriate command to end your session.

```
PTS
61235421+ (Return)
GA
/BATCH

LLLL
```

To upload addresses and messages, you must first create a file on disk. Be certain to type them just as you would if creating them online. Since you will not receive a GA prompt, you must type "/TEXT" on an empty line following each "+" sign. To end each message, type "LLLL" after all messages except the final one. After the final message, type "MMMM."

To upload the file, logon the system in the usual way. At the PTS response, type "/BATCH," and press Return. Then take whatever action is required by your software to transmit the file.

```
PTS
/BATCH (Return)
```

EasyLink informs you of the number of messages sent, and prompts with a new session number, date and time, and PTS response.

The Electric Mailbox

READING MAIL

As an EasyLink subscriber, you are provided an "intelligent" mailbox where messages are delivered and stored until read. If a mailbox is not accessed for more than 10 days, a paper copy of all unread messages is automatically mailed to you in the form of a Mailgram. Since there is a charge of 50 cents for each message delivered, it's a good idea to check your mailbox frequently.

Messages are received in your mailbox, and once read, they are placed in you Mailbox Hold File for an additional three days. At midnight on the third day they are automatically deleted from the system. EasyLink does not offer a filing system, so in order to retain a copy of the message you must save it to disk or print it out.

Three commands can be used with your EasyLink mailbox: /Scan, /Read, and /MBX. These commands are always preceded by a slash ("/"). Let's review each of them.

/Scan. Displays a list of all messages. Messages must be scanned before they can be read. By scanning the list, you can determine the message number of the message you want to read. Messages cannot be read using this command. Type "/SCAN HOLD" to view the contents of your Mailbox Hold File.

EasyLink does not automatically pause at the end of each screenful. To ask the system to pause after a specified number of lines, type the /Scan command followed by the number of lines you want displayed. For example, "/SCAN 10" pauses after every 10 lines. Type "/SCAN CRT" to display 22 lines at a time.

/Read. Messages in your mailbox can be read in one of two ways: scan your mailbox and read messages selectively; or remove all messages at one time.

The /Read command can be used only after the mailbox has been scanned since the command must have one or more message numbers entered after it. Type "/READ" followed by the number of the message you want to view. Message numbers may be entered singly, in a series, by a range of numbers, or a combination of all of these. Separate each number or range of numbers by a comma, but no space. To read all the messages in your mailbox type "/READ ALL."

    /READ 1,4,7-10

EasyLink

/MBX. Use the /MBX command to read all the messages in your mailbox. Unlike the /Read All command that is used after scanning, /MBX allows you to read all your messages without first viewing the Scan list. The messages scroll up the screen continuously until the last message is received. To prevent losing text by messages scrolling off the screen, type "/MBX CRT." This command displays 22 lines of text at a time. You can also type "/MBX" followed by any number from 10-99 to request a screen display suited to your equipment requirements. After the requested number of lines are displayed, you receive the prompt "Press (Return) To Continue." Do not use these command with /Read. An alternate to the CRT command is to use CTRL S to stop the scroll and CTRL Q to begin the scroll again.

OTHER EASYLINK SERVICES

EasyLink provides a number of additional services to its subscribers. They are briefly discussed here. For detailed information, consult your EasyLink User Manual.

Sending Telex Messages

Telex I and Telex II (also called TWX) are Western Union's domestic Telex systems. Messages can be sent only within the United States, Canada, and Mexico. Messages also can be sent to non-Western Union domestic Telex numbers, other than Telex I and II, including RCA, FTCC, Graphnet, ITT, and TRT.

A Telex message cannot exceed 200,000 characters and the line width must not exceed 68 characters. When sending Telex I messages, type all characters in capital letters. Certain symbols are not permitted.

To send a Telex message, type the Telex number, an answerback, if one is used, a "+," and press Return. Then type your message, press Return, and type the appropriate sending option, "LLLL" or "MMMM," to send the message.

Using WorldWide Telex (/WUW)

Over 1.5 million subscribers in 154 countries can be reached using WorldWide Telex. These messages cannot be sent to United States, Canadian, or Mexican Telex numbers. It is not necessary for an attendent to be present for a message to be received, because as long as the Telex terminal is left on, messages are received and printed any time of the day or night.

The Electric Mailbox

WorldWide Telex messages can contain up to 200,000 characters, but must not exceed 68 characters per line. And there are limitations on special characters. The command "/WUW" is used send a WorldWide Telex message. The Telex number is always preceded by a three digit "country code" that identifies the destination country for EasyLink. A complete list of codes can be found in the user manual or obtained online by typing "/HELP COUNTRY" at the PTS response.

WorldWide Telex messages are priced by the minute and have a one minute minimum. The rates are different for each country. To obtain specific rate information, enter /HELP COUNTRY at the PTS response.

To send a message, type "/WUW," a space, the 3-digit country code and Telex number, an answerback (if one is used), a "+," and press Return.

/WUW (space) (3 digit country code) (telexnumber) (answerback) +

/WUW 7426745231(6745XYZ)+

At the GA prompt, type the text of the message, press Return, and type the sending option (LLLL or MMMM.)

Sending Mailgrams (/ZIP)

Mailgrams are inexpensive overnight letters sent electronically by satellite to post office receiving centers and then delivered in the next day's mail. Currently, Mailgrams can be sent anywhere in the United States, Canada, and Puerto Rico, although the service is scheduled to begin in the United Kingdom in the near future.

The cost for Mailgrams is $3.45 for the first half page and 50 cents for each additional half page. A $1.15 surcharge is made for Mailgrams delivered to Canada.

Mailgrams can consist of up to 15,000 characters. A full page can have a total of 52 lines, but no more than 68 characters are permitted on each line. An electronic page consists of 2,500 characters, including control and format characters. Thus, Mailgrams are restricted to approximately 4 pages.

To send a Mailgram, type "/ZIP" at the PTS response and press Return. Next, type the name and address of your correspondent. A Mailgram may have up to 5 address lines, but each line must be no longer than 44

EasyLink

characters and contain no punctuation. Press Return after each address line. To signal the end of the address, type a "+" sign and press Return.

When sending a Mailgram to Canada, enter only the province on the address line; "Canada" is not required. A list of abbreviations for Canadian provinces is located in the user manual. For messages being sent to Puerto Rico, enter PR in place of the state abbreviation.

Normally, your return address is the one associated with your EasyLink ID. However, you may request that a different return address appear on the Mailgram. A custom return address is entered immediately following the recipient's address and can contain up to 3 address lines consisting of no more than 30 characters per line. Enclose the entire address in parentheses. When a custom address is used, do not type a "+" following the recipient's address. Type it following the closing parenthesis.

```
    PTS
    /ZIP
    Robert Rondell
    8645 Palm Dr
    North Rockridge CA 90345
    (Joan Blossom
    631 Mountain Pass
    Brockfield MA 05846)+
```

Sending Telegrams (/PMS)

Telegrams are messages delivered to the recipient over the phone by a Western Union operator. Usually, they are delivered within 4 hours from the time sent. A paper copy of the message can be requested. Telegrams can be sent anywhere in the United States, except Hawaii, and to Canada. Telegrams are charged at the rate of five cents per word plus a $2.75 service charge. Western Union defines a word as consisting of 6 characters. (A space, line feed, and return each count as one character.) A telegram can contain up to 15,000 characters with a line width of no more than 68 characters.

To send a telegram, type "/PMS," a space, the name of the recipient, and press Return. On the following lines, type the address. The address can consist of up to 5 lines, but must contain no punctuation. Type the city, state, and zip code on a single line. Do not enter "Canada" for Canadian Telegrams; use the province abbreviation instead. Type a "+" following the last character in the zip code to signify the end of the address entry and press Return.

The Electric Mailbox

At the GA prompt, type the message. If the telephone number of the recipient is known, type it as the first line of the message, press Return, and complete the message. Press Return again to end the message and type the appropriate sending option, MMMM or LLLL.

    PTS
    /PMS Sally Jones
    4856 Basset Rd
    West George CT 00124+

    GA
    TELEPHONE: 204-555-9933

Sending Cablegrams (/INT)

Cablegrams are international or overseas Telegrams that can be delivered by telephone or mail depending on the country to which they are sent. Use Cablegrams to send overnight messages to all foreign countries (except Canada), and to Hawaii.

Rates vary for Cablegrams depending on where they are sent. To obtain the correct fee, type "/HELP" at the PTS response followed by the country to which the Cablegram is to be sent.

    /HELP NETHERLANDS

Cablegram messages can contain up to 15,000 characters including the address and text. Text must be typed in upper case and there are restrictions on special characters.

To send a Cablegram, type "/INT," a space, the recipient's name, and press Return. Type the address. The address can be up to 6 lines in length, but commas and other punctuation are not permitted. The city and country are entered on the final address line. Enclose the country name in parentheses followed by a "+" and press Return. At the GA prompt, type your message, press Return, and enter the appropriate Send option.

    /INT William Davis
    645 Rue Leon
    Paris (France)+

EasyLink

Sending InfoCom Station Letters (/ICS)

An InfoCom Station is a private Telex type communications network. Private wire customers and InfoMaster switching facilities are used to send internal messages, thus avoiding private message-switching system costs. You can only send messages to these stations if the station is willing to accept them. InfoCom Station messages can be up to 200,000 characters in length with a line width of no more than 68 characters.

To send a message to an InfoCom Station, type "/ICS," a space, the routing code for the station, a "+," and press Return. Type the message in the usual way, press Return, and type the appropriate Send option.

    /ICS KDGSFEV+

Sending Computer Letters (/CLS)

Computer Letter Service provides computer-generated letters that are mailed by the United States Post Office anywhere in the 50 United States and Canada. Delivery is usually made within three business days. Computer letters can contain up to 371 lines with a maximum of 68 characters per line (approximately 5 1/2 pages). Only one letter can be sent at a time. These letters cannot be sent with notification of delivery or by priority delivery. Additionally, attention lines and alternate addresses are not permitted. The cost of a computer letter is $1.50 for the first page and 50 cents for each additional page.

Computer letters are sent initially to a special EasyLink mailbox for processing. To send a Computer Letter, type mailbox number "62900396+" and press Return. Following the GA prompt, type "/CLS" and press Return. You are now ready to address the letter.

An address can be up to 6 lines and contain no more than 40 characters per line, and no punctuation. The city, state (province for Canada), and zip code are typed on the last address line. Press Return following the address. Computer letters do not require a "+" sign.

    GA
    /CLS
    Marshall Roberts
    45 Newhardt Lane
    Bonnieville TX 78352
    /TEXT

The Electric Mailbox

Unlike other EasyLink messages, you do not receive the GA prompt. Instead, type "/TEXT" and press Return. Type the message, and on the line following the text type "/END" and press Return. End the session by typing the sending option of your choice (LLLL or MMMM).

Sending Overseas Priority Letters (/OPL)

An Overseas Priority Letter is a computer letter that can only be sent to the United Kingdom. It is faster than air mail and less expensive than a cablegram. Only one letter can be sent at a time. These letters are usually delivered by the British Post Office within two business days of receipt. Notification of Delivery, Priority Delivery, and Attention Lines are not permitted. Check the user manual for other restrictions. Type "/RATES" at the PTS response to obtain costs for this service.

Overseas Priority Letters are prepared in the same way as Computer Letters with the following exceptions: the /OPL command is used to alert EasyLink that you want to send an Overseas Priority Letter; the letter is limited to 363 lines; and the country name is placed in parentheses followed by the postal code for the address, if known.

London (England) BR3C 8BL

Sending Express Documents (/OVNT)

EasyLink provides an easy way to send high quality documents printed in letter quality on bond paper: the Express Document Service. Delivery is made overnight by the DHL Worldwide Express service. Overnight delivery is available in thousands of towns and cities nationwide and will soon be available in selected international locations. Delivery is guaranteed by noon, local time, the following day. To obtain a list of available locations, type "/HELP STATE" at the PTS response.

Do not exceed 65 characters per line or 54 lines per page. If you do, EasyLink will reformat the message to fall within these margins. If you want less than 54 lines per page, type "/PAGE" on a new line to indicate a page break. This command enables you, for example, to prepare a title page for your document. You can use the /PAGE command as often as you like. Use the space bar to enter columns in the text.

To send an Express Document, type "/OVNT" following the PTS prompt and press Return. At the beginning of the next line, type the name and address just as if preparing an envelope for a normal letter, but do not use any punctuation. Type a "+" sign and press Return at the end of the address. When the GA prompt is displayed, type the text of the message and end it with either MMMM or LLLL.

A nice feature of the express document service is that you can provide special handling instructions for the courier. The command "/DELV" is used to request this service. Type the instructions on the line following the last line of the address. Instructions can consist of up to 60 characters, but do not use commas. When this feature is used, type the "+" following the delivery instructions rather than after the last line of the address. The instructions do not appear on the prepared document, only on the courier's delivery address sheet.

```
PTS
/2HOURJames Jarvis
President
ABT Communications
86 North Ruggerford Rd
Boston MA 01632
/DELV Leave with president's secretary+
```

LOGOFF PROCEDURE

Type "/QUIT" and press Return following a PTS response to disconnect from EasyLink.

CONCLUSION

EasyLink is a powerful telecommunications system that not only allows the user to communicate with other EasyLink users, but literally puts the world at his fingertips. The experienced user is not hampered by menus, but has them close at hand should the need arise. The novice can enter the prompt mode at sign on and use it until he feels familiar enough with the system to operate without them. And help is always available either online or from the Customer Care Center. EasyLink is probably better suited to the high volume user, since the cost is relatively high in comparison with other services. High volume users can take advantage of substantial discounts.

InfoPlex

InfoPlex
CompuServe Information Service, Inc.
5000 Arlington Centre Boulevard
Columbus, OH 43220
800-848-8199, 800-848-8990
614-457-8600

InfoPlex is CompuServe's corporate electronic mail system. It was introduced to the market in late 1978. Sales and service facilities are located in more than 30 major US cities, providing marketing and technical support to large commercial clients. CompuServe provides information services to small and large organizations, both in the private and public sectors.

In addition to its InfoPlex electronic mail service, CompuServe provides access to information on financial planning and analysis, database management, investment banking, management services, research and development, and engineering. InfoPlex customers can have full access to the CompuServe Information Service (CIS) that is described in the EasyPlex chapter of this book. Messages can be exchanged between the two services.

Each subscribing corporation selects an InfoPlex administrator. The administrator trains new users, sets up individual mailbox codes, and integrates InfoPlex into the day-to-day operation of the company.

HOW TO SUBSCRIBE

InfoPlex is sold exclusively through CompuServe's 31 marketing offices nationwide by a team of account executives. Check the yellow pages of

InfoPlex

your telephone directory for a CompuServe office near you, or call CompuServe at 614-457-8600.

RATES

The normal sign-up fee of $89.95 includes two free hours of connect time and system documentation. There is a monthly minimum usage of $500 per month after the first four months of service. Access charges are $12.50 per hour during prime time and $6.00 per hour during non-prime time (evenings, weekends and holidays) when accessing the service from over 200 cities with local CompuServe numbers. The service may also be reached through Tymnet, Telenet and WATS numbers, though additional communications surcharges apply. There are also per message charges based on the number of characters in your message.

Rates for InfoPlex are subject to change. Current rates may be obtained from an account executive in one of CompuServe's marketing offices. Usage of InfoPlex can be billed by transaction or by flat usage, and other billing structures can be developed to meet the needs of the client.

ACCESSING THE NETWORK

InfoPlex may be accessed through CompuServe's own network, Tymnet, Telenet, or Datapac. There is a communication surcharge made for these services. Refer to the appendices for network access numbers.

CONNECTION PROCEDURES

The procedure for connecting to CompuServe depends on the network you are using. Before you sign-on, you will need your user ID, password, address, and access code. Your address and access code are assigned by your InfoPlex administrator. Once connection is made, you may proceed to the "Logon Procedures" described in the next section.

CompuServe. Dial your local access number and wait for the carrier tone. Then press CTRL C. If you receive the "Host name" prompt, type "CIS" or a code given to you by your InfoPlex administrator, and press Return. You may proceed to logon.

Tymnet. Dial your local access number and at the "please type your terminal identifier" prompt, type "A." Tymnet then displays the prompt "please log in." Type "CPS01" or "CIS02." If you are using a half duplex terminal, precede your entry with a period, for example ".CPS01." You are now connected and may proceed to logon.

The Electric Mailbox

Telenet. Dial your local access number and when connection is made, press Return twice. Telenet responds with the "Terminal=" prompt. Type "D1" and press Return. Then at the "@" symbol, type "C 202202" or "C 614227." If you receive the "Host name" prompt, type "CIS" or "CPS." You are now connected and may logon.

Datapac. Dial your local Datapac number. When you hear the tone, type the service request signal. For 300 baud, type a period (".") and press Return. For 1200 baud type two periods ("..") and press Return. The periods will not show on your screen. Datapac responds with "DATAPAC." You must now indicate whether you want to connect via Tymnet or Telenet by typing one of the following call request codes and pressing Return:

Tymnet	Telenet
P 1 3106,CPS01	1311020200202
P 1 3106,CIS02	1311061400227

if your connection is successful, you will receive the "User ID" prompt and may proceed to logon. If you should receive the "Host name" prompt, type "CIS" or "CPS."

LOGON PROCEDURES

Logging on to InfoPlex is accomplished in four steps. Your first prompt is for your User ID. This is the two-part number supplied by CompuServe for your organization. Type your ID and press Return. Then at the Password prompt, type your password and press Return again. The Address prompt is displayed. Your address is comparable to the name on a mailbox and is supplied by your organization's InfoPlex administrator. Enter your assigned address and press Return to receive the Code prompt. The Access Code serves as the key to your mailbox. A correct entry opens your mailbox and generates a banner indicating the date and time of your logon. You will be informed if you have messages waiting. A typical logon looks like this:

```
User ID: 12345,6789 (Return)
Password: XXXX.XXXX (Return)
Address? B.SIVART (Return)
Code? OXOXOX (Return)
InfoPlex 1E(67) -- ready at 10:36 EST 24-Dec-86 on M83PST
6 Messages pending
```

InfoPlex

As a security measure, your password and access code will not be displayed on the screen.

For certain types of terminals, you need to enter a terminal designator immediately following the User ID. For example, if your terminal is a Selectric, type your ID followed by an asterisk (*), the terminal identifier ("T" in this case), and press Return.

    User ID: 12345,6789*T

See the CompuServe appendix for for a list of terminal designators.

InfoPlex provides an additional level of security, a Confidential Code. This code is useful if many users have access to the same mailbox. When you use the code, only you can read a confidential message. The code must be entered at logon in place of the Access Code.

USING INFOPLEX

InfoPlex is a command driven email service. Commands may be issued using their full name or, in most cases, abbreviated to the first three letters of the command. Commands are always preceded by a slash (/) and followed by pressing Return. The first time a command is introduced in this chapter, the abbreviation will be shown in parentheses. If you make an error while typing the command, simply use the Backspace key to move to the place the error was made and retype the command.

Some commands require more information, such as a message number or a date. In this case, type the command, a space, and then the additional information. In some situations, the message "[Ready]" appears on your screen. This indicates that you can proceed with your session. In other cases, the message "[Done]" is displayed, indicating that your command has been carried out and that you may enter another command.

InfoPlex performs an automatic disconnect whenever your terminal remains inactive for a specified number of minutes. This would occur in situations where no information has been entered, received or sent in the specified time. It is a protection to you in case you leave your terminal unattended.

Getting Help

Help is available online by typing the Help command at any prompt. This displays a list of all available topics followed by a prompt for a specific

The Electric Mailbox

topic. Some topics offer a list of help files available for subtopics. Or you may type "HELP" followed by a topic name to receive information about that topic. To end the Help display, press Return at a New Topic prompt. Of course, help is always available by calling Customer Service at 800-848-8990 from outside Ohio and in the contiguous US, or 614-457-8650 from Ohio or outside the contiguous US.

Special Control Characters

The following control characters may be used during your InfoPlex session.

CTRL A	Stops display of output at the end of the current line.
CTRL Q	Stops display at the current character.
CTRL C	Stops display and terminates the current command.
CTRL H	Causes the cursor to backspace.
CTRL I	Functions like a tab stop on a typewriter. Default settings are at positions 9, 17, 25, 33, 41, etc. Each tab stop is 8 spaces beyond the previous one.
CTRL P	Terminates processing of an InfoPlex command.
CTRL U	Deletes the line you are working on.
CTRL W	Resumes the display of text at the point it was interrupted by a CTRL A or CTRL Q.

An interesting feature of the InfoPlex system is that in the event of a power shortage or idle time-out, your workspace, the area in which you create messages, will be saved automatically by the system. When you logon after such an event, you will be informed that a backup file exists, and asked if you want to use it or delete it. If you type "USE," the file will be placed in your workspace, just as if nothing had happened.

With the preliminaries out of the way, let's get started with our InfoPlex session and learn how to compose and send a letter.

SENDING A LETTER

The Compose (Com) command is used to set up a workspace in which to prepare a message. You may edit messages while in Compose mode by using the control characters discussed earlier, or by typing "/EDIT" to use any of the editing commands available in InfoPlex's basic or advanced editing function. (See Chapter 4, Editing Your Messages, in the InfoPlex Users Guide.)

InfoPlex

Do not type more than 130 characters on each line, and end each line by pressing Return. A message may be up to 50,000 characters in length. To create a blank line, press Return on a line by itself.

To prepare a letter, type "/COMPOSE" and press Return. Then type your message just as you want it to appear to your recipient, and when it is complete, type "/SEND" (Sen) on a line by itself followed by the address of your correspondent, and press Return. Before you enter the Send command, you may delete the letter and start over again simply by issuing a new Compose command on a line by itself. Following the Send command, you will be prompted for a subject that may contain up to 80 characters. Type the subject and Press Return. The message number, and the date and time it was sent are displayed.

>   /COMPOSE
>   [Ready]
>
>   We would like to propose an experiment that could
>   revolutionize communications in this country: rapid
>   mail delivery by Pony Express. We promise to deliver mail
>   posted at St. Joseph to Sacramento within 10 days.
>   We eagerly anticipate the opportunity to discuss this
>   with you further.
>
>   Sincerely,
>   Russell, Majors and Waddell
>
>   /SEND WHITEHOUSE
>   Subj?: Express Mail Delivery
>
>   Message 987-654 sent at 12:34 EST 03-Apr-60

Since the Send command does not delete the letter from your workspace, you may issue another Send command if you forgot to include other intended recipients. Also, the message can be edited after sending it, permitting you to send different versions to other IDs.

Sending Options

Options available for use with the Send command follow.

DIS   *Distribution.* Appends a list of the addresses to which the message will be sent.

The Electric Mailbox

CON *Confidential.* Sends a message that can only be read if the recipient logged on using his Confidential Access Code.

PER *Personal.* Prevents a personal message from being forwarded by the recipient, or from being copied into the recipient's workspace.

PRI *Priority.* Lists the message before non-priority messages in the recipient's mailbox.

REC *Receipt.* Sends you an automatic confirmation notice when the recipient has read your message.

REL *Release:mm-dd-yy.* Enables you to establish a specific date for delivery of your message. You are notified when delivery is made.

REQ *Require.* Requires a reply from the recipient. The message remains in the recipient's mailbox until a response is sent.

These options are entered immediately following the Send command and multiple options may be entered on the same line. For example, to send a personal message to A.Baldwin with a request for reply, you would make the following entry:

/SEND/REQ/PER/CON A.BALDWIN

In special situations, there are two commands that can be used instead of the Send command.

FOR *Forward.* Sends a received message to another individual. Type "/FORWARD" followed by the message number.

/FORWARD 986-123

Comments may be entered at the top of the message by appending the Comments (Com) option to the Forward command; or to the bottom of the message by appending the Comments (Bot) option following the Forward command.

/FORWARD/COMMENTS 986-123
/FORWARD/COMMENTS:(BOT) 986-123

RES *Respond.* Prompts for a response to a message sent to you with the Require option.

InfoPlex

InfoPlex also provides commands that may be used on messages that have already been sent.

CON *Confirm.* Verifies whether a message you sent has been received. The command may be followed by a message number or an address.

> /CONFIRM 246-810
> /CONFIRM D.BAND

> If the system finds the message, you know that it is in the mailbox of the recipient, but has not been read. If the message is not found, you know that it has been received and read.

ERA *Erase.* Deletes a message from the recipient's mailbox. This command only works if the message has not been read. Infoplex is one of the few services that offers the ability to manipulate mail after it has been sent.

UPLOADING FILES TO INFOPLEX

InfoPlex permits transfer of normal text files as well as binary files such as spreadsheets. In order to transfer files between your computer and InfoPlex, you must specify the protocol to be used. InfoPlex supports the following protocols:

> CompuServe B protocol
> XMODEM (MODEM7) protocol
> DC2/DC4 Capture (buffer capture) protocol
> No protocol

You may specify the protocol by issuing a terminal protocaol designator (see CompuServe appendix), or by including a protocol specification in the file transfer command. Use the Upload (UPL) command to initiate a transfer. If Upload is entered followed by pressing Return, you will be prompted for the protocol, file type, disk drive, and filename. But it is easier to include all the information on the command line. The following command would upload the ASCII file "ACCT.FIG" located in the B Drive, and use the XMODEM protocol.

> /UPLOAD/XMODEM/ASC B:ACCT.FIG

At this point, consult your communications software for appropriate actions to send the file. When the transfer is complete, you are returned to the

InfoPlex Command mode. (Consult Chapter 5 of the User Guide for detailed information.)

READING MAIL

There are four commands and numerous options that may be used to read and handle your incoming mail: Scan, Receive, Save, and Delete. When you first logon the system, you are informed if you have mail waiting. At this point you may use the Scan (Sca) command to check the contents of your mailbox; or you may issue the Read (Rea) or Read All (Rea All) command to immediately see all messages. The "all" does not have to be entered since InfoPlex assumes that Read refers to all messages. Now, if you have confidential messages in your mailbox, they will not be displayed unless you entered your Confidential Access Code when you logged on. First, let's take a look at how to use the Scan command.

The Scan command displays a list of all messages in your mailbox and provides a brief description of each, including the subject, the name of the originator, the date and time the message was sent, and the length of the message by the number of characters. Additionally, messages sent with the tags Confidential, Response Required, Receipt, or Priority are identified. Each message in the scan table has a sequence number based on its order of receipt. The exception is a Priority letter that is always placed at the top of the list. Additionally, the system assigns each letter a unique message number. Confidential messages will be listed, even if you have not logged on using your Confidential Access Code. You just won't be able to read them. Shortly, we will see how the sequence numbers and message numbers are used when issuing some commands.

The following options are available for use with the Scan command. The option is entered immediately following the command, without spaces. For example,"/SCAN/CON."

CON *Confidential.* Lists only confidential messages.

PER *Personal.* Lists only messages marked as personal.

PRI *Priority.* Lists only priority messages.

TOD *Today.* Lists messages sent within the last 24 hours.

FRO *From:(address).* Lists messages from an address you specify. More than one address may be entered following the command. And

InfoPlex

broadcast codes may also be used with this option. Enclose the address in parentheses.

SIN *Since:(date)*. Lists only messages sent on or after the date identified in the command. Enclose the date in parentheses.

BEF *Before:(date)*. Lists messages sent prior to the date specified in the command. Enclose the date in parentheses.

DAT *Date:(date)*. Lists messages received on the date you specify. Enclose the date in parentheses.

NUM Number. Renumbers the remaining messages in the Scan table. Use this option only after you have deleted some of the messages from your mailbox.

Multiple options may be entered on the same line. For example, to Scan all private messages received from J.Jones since December 1, 1986, enter this command:

/SCAN/SIN:(12-01-86)/FRO:(J.JONES)/PRI

The Receive (Rec) command continuously displays messages in your mailbox. Messages may be received all at once, one at a time, or selectively.

/REC ALL
/REC 1,5-9,12,14

The same options available for listing messages using the Scan command are available for displaying messages when using the Receive command. The exception is the Number option that can only be used with Scan. The following special options may be used with the Receive command.

FEE *Feed*. Performs a form feed by placing blank lines between messages or displaying each new message at the top of the page or screen.

PAU *Pause*. Pauses after printing each message. Press Return to print the next message.

The Read command displays incoming messages one at a time, and then asks what action you want to take on them. Any option used with the Receive command may be used with this command. Here are some of the actions that can be taken on a just-read message.

The Electric Mailbox

Delete (Del). Erases one or more received messages. Remember, all messages that have been received but not saved are automatically deleted after you terminate a session. Thus, this command is of little value unless you want to reduce you messages to a more manageable group.

Edit. Permits you to revise the contents of the message using the InfoPlex Edit function.

File. Places a copy of the message in your filing cabinet. You will be asked for a topic. The scan description of the message is then displayed and you are prompted for further action.

Forward. Places you in the edit mode and prompts you for comments. A message marked personal may not be forwarded.

Next (N). Advances to the next message.

Reply. Prompts you for a reply and places you in the Edit mode.

Reread. Redisplays the entire contents of the just-read message.

Save (Sav). Retains one or more received messages in your mailbox. To save all messages, type "/SAVE ALL." To save specified messages, enter the Save command followed by a message number. All saved messages remain in your mailbox until they are received again.

Use. Places the message into your workspace where it can be edited and sent to another mailbox. This command may not be used with any message marked personal.

You will find other useful commands and options as you review your user's guide. But these should give you an idea of the ability of InfoPlex to help you manage your day-to-day correspondence.

SENDING A TELEX MESSAGE

InfoPlex messages may be sent to a telex or TWX machine anywhere in the world. However, to use this service you must be authorized by your administrator. Check with your InfoPlex administrator to determine if you are eligible for the service.

To send a telex message, you must know the telex number and answerback (optional) of your correspondent. Your telex command must be formatted in this way: "/SEND >TLX <number> <answerback>"

InfoPlex

The answerback is optional, but is an excellent way to verify that your message is being received by the right machine. A telex message may also be sent to mailboxes at the same time. Type the mailbox number on the same line as the telex number.

/SEND SMITH, >TLX 1234567890 ABTCOM, ROBERTS

After the telex message is sent, you receive a confirmation in your mailbox verifying whether it was successfully sent. If the message could not be delivered, you receive a copy so that you can resend it at a later time.

OTHER SERVICES

CompuServe and MCI have announced an interconnection between their electronic mail systems. InfoPlex customers can send electronic mail to MCI Mail customers and vice versa. In addition, EasyPlex customers can send and receive electronic mail with MCI customers and with InfoPlex customers.

Sending a Message to MCI Mail

To send a letter to an MCI Mail subscriber, after the Send command, type a space, a greater-than symbol (>) and "MCIMAIL:" followed by the username or ID of the MCI Mail subscriber to whom the message is being sent.

/SEND >MCIMAIL:123-4567

or

/SEND >JCAESAR

LOGOFF PROCEDURES

There are several commands that may be used to logoff InfoPlex.

BYE. Use this command when you want to terminate your session with no immediate re-access. You will be notified of the date and time of your logoff.

EXIT. Use this command if you are a CompuServe Interchange user and originally accessed InfoPlex from an Interchange menu. You will be returned to that menu.

The Electric Mailbox

LOG. Use this command if you intend to logoff under one user name and logon with another User ID assigned to your organization.

CODE. Use this command to end the session for the current address and begin a new session for another address without signing off CompuServe.

Remember, you will be automatically disconnected if you fail to use the system for a specified period of time.

CONCLUSION

InfoPlex is easy to learn and use. Because of the monthly minimum usage fees, it is best suited for large companies that intend to send a considerable volume of messages. Though it offers no direct paper mail option, it is versatile enough for most firms needing instant internal communications. Training courses are held each month in each of CompuServe's local offices.

Telemail

GTE Telemail
GTE Telenet Communications Corporation
12490 Sunrise Valley Drive
Reston, VA 22096
703-689-6000

GTE Telemail was inaugurated in 1975 and now boasts public data network access from 400 cities in the United States and more than 70 overseas locations. In addition to providing its users the ability to send and receive Telemail messages to and from the system's users, subscribers can send and receive telex messages over Telemail, communicate with Telemail users in other countries, create and use "electronic" forms, and communicate via bulletin boards.

The TelemailXpress service was introduced in late 1984 and provides a hard copy delivery service to any place in the world, even if the recipient is not served by electronic mail. Two options are available: laser-printer business quality letters processed in one of 14 locations across the country and mailed first-class through the post office; and overnight delivery messages printed and sent through express mail for next day delivery. Printed matter, such as purchase orders and invoices, can be delivered using TelemailXpress service.

Telemail is straightforward, easy to learn, and easy to use. Instructions are presented in a flipchart format, making location of needed information easy to find. A user's guide is also available and provides in-depth instructional material. Telemail offers an online tutorial to help you become familiar with the basic Telemail concepts. Just type "COMPOSE TRAINER" at the Command prompt.

HOW TO SUBSCRIBE

Make application by calling GTE Telenet Communications Corporation at 800-TELENET or 800-368-4215.

RATES

There is no initial fee for signing up for the service. However, subscriber's accounts are charged $140.00 per month, and there is a $500.00 per month minimum usage fee. The rate for online time varies with the time of day the service is used and the user's geographic location.

Business rates apply Monday through Friday, 7 a.m. to 6 p.m. From the continental US, the business rate is $14.00 per hour (public dial) and $11.00 per hour (private dial/dedicated access). Rates for Hawaii are $20.00 per hour, and international points $14.00 per hour. A charge of $32.00 per hour is made for WATS calls. A PTT network charge is added to the Telemail international rate. Surcharges include $28.00 per hour for Alaska, $23.00 per hour for Canada, $30.00 per hour for Mexico, and $25.00 per hour for Puerto Rico.

Off-peak rates apply Monday through Friday, 6 p.m to 9 p.m., and from 7 a.m to 9 p.m. on Saturdays, Sundays, and major holidays. From the continental US, the off-peak rate is $7.00 per hour (public or private dial), $12.00 per hour from Hawaii, $14.00 from international points, and $21.00 per hour for WATS calls.

Night-time rates apply 9 p.m to 7 a.m. everyday. From the continental US, the rate is $4.00 per hour (public or private dial), $10.50 per hour from Hawaii, $14.00 per hour from international points, and $10.00 per hour for WATS calls.

Additional charges are made for storage, broadcast delivery, auto delivery/direct delivery, and special reports.

Telemail supports baud rates 300 through 2400.

ACCESSING THE NETWORK

GTE Telemail is accessed through the Telenet communications network. See the appendices for a list of telephone access numbers.

Telemail

CONNECTION PROCEDURES

To connect to Telenet, dial your local Telenet telephone number. When you get the carrier tone, you know you have made connection. Press Return twice and begin your logon. At the "Terminal=" prompt, type your terminal ID. Telemail's booklet, "How To Use Telenet," offers a list of terminal codes for many terminal models. Or you can call customer service for help. Most computer terminals can use the "D1" terminal ID.

LOGON PROCEDURES

Next, you are prompted with the "@" symbol, which is Telenet's way of asking what you want to do. Type "MAIL" to access Telemail. At the Username prompt, type your username and press Return. When another person on the system has your same username, you are asked for further information, such as the name of the organization you represent. The Organization prompt can be avoided by typing your username and the name of your organization on the same line, separated by a slash ("/"). You are prompted for your password. Type your password and press Return. As a security measure, the password will not print on the screen.

>   Username? SSMITH/ABC.PROD (Return)
>   Password? XYXYXY (Return)

The welcome banner is displayed, and if new messages have been posted to a monitored bulletin board, you receive a message to check those boards. This is followed by a scan list of all new messages received since you last logged on. Then you will see the Command prompt.

USING TELEMAIL

Every Telemail user is provided a *catalog* that serves as a personal work and storage area. The catalog has two compartments: a mailbox to hold messages until they are read, and a storage area to hold messages after they have been read or have been created and saved in files. As a user, you also have access to bulletin boards that are used in much the same way an organization uses bulletin boards to post messages and announcements of interest to all. Some boards are available to any subscriber on the system; others are maintained by individual organizations and accessible only by individuals within the organization.

The Electric Mailbox

Telemail Prompts

Telemail communicates with you through a series of prompts and system messages. System messages provide additional information or alert you to errors. A complete list of system messages is located in the User's Guide, Appendix D. Each message is defined and followed by the action you should take.

Getting Help

Telemail offers online assistance, as well as the ability to contact system personnel regarding Telemail-related problems or questions. To receive help online, type a question mark ("?") or "HELP" followed by a command name at any Command? or Action? prompt. You receive a brief summary of the command and a list of available options. For example, type "? COMPOSE" to receive help in creating messages to other Telemail users. You can also get help from your organization's Administrative Manager (ADMIN). The ADMIN is responsible for maintaining the way Telemail is used in your organization and serves as a liaison with GTE Telenet. Customer service is available by calling 703-442-1900 or 800-368-3407.

Special Control Keys

Knowledge of the following special function keys will make your Telemail experience more enjoyable.

CTRL H	Deletes a character
CTRL W	Deletes a word
CTRL X	Deletes a line
CTRL R	Redisplays the most recent line entered
CTRL S	Stops display and freezes output
CTRL Q	Resumes display and continues output

The primary commands used in the Telemail system are: Scan, Read, Compose, Answer, Forward, Purge, Unpurge, and Check. Each command serves as a base from which a variety of options can be chosen to do a specific task. Some options are common to several commands; others are peculiar to a single one. Commands may be entered in upper- or lowercase and some can be abbreviated. The abbreviations are noted in parentheses the first time the command is used below. Let's begin our Telemail adventure by learning how to send a letter.

Telemail

SENDING A LETTER

"Compose" (C or COM) is the command used to prepare messages online to send to other Telemail users. To send a letter, type "COMPOSE" at the Command prompt. The system will prompt for the envelope information (To:, CC:, and Subj:). First enter the recipient's name at the "To:" prompt and press Return. More than one username can be typed on the same line, but each name must be separated by a comma. At the CC: prompt enter the names of those to whom you want to send courtesy copies and press Return. Or just type Return if you do not want to use this option. You can also send a blind copy by entering "(BCC)" after the username. (BCC) may be used at either the To: or CC: prompt. There is a 132 character limit per line for both these prompts.

    To: Dopey,Sneezy,Sleepy(BCC)
    CC: Grumpy,Bashful,Happy,Doc(BCC)

The Subj: prompt is displayed next. The subject can contain up to 132 characters. If you do not want to make an entry, press Return to receive the Text: prompt. Each line of text can contain up to 132 characters, and there are no restrictions on the length of the message. To end the message, type a period (".") on a line by itself and then press Return. This generates the Send prompt.

    Subj: Dinner plans changed

    Text:

    Don't hold dinner for me. Got a heavy date with Prince
    tonight. I made you a fresh apple pie. Help yourselves.
    See you in the morning...maybe.

    Hugs,
    Snow
    .

    Send?

The Send prompt is asking whether you want to send the message. You have four options for handling the composed message: send, delete, or edit it, or save it as a workspace. An entry of "Y" means that you want to send the message without any modification. Confirmation is received indicating the date and time the message was posted and the system message number (SMN) assigned by Telemail. Once a message is sent, there is no way to retrieve it. If you decide that you want to modify the message or add delivery options, type "N" at the send prompt.

Send Options

Telemail offers several special delivery options in addition to the Send option.

(PRI) (Private). Used when you want to ensure that only the intended person reads the message. Before the recipient can read the message, he must enter his personal ID.

(REC) (Receipt). Used when you want to receive a return receipt. You are notified in the scan table of the date and time the message is read by the recipient.

(REG) (Registered). A registered message must be acknowledged by the recipient before it can be read. You are notified of the acknowledgement by a short message sent to your mailbox indicating the date and time it was read.

(URG) (Urgent). Used when you have an important message to send. Messages sent with an Urgent tag are placed at the top of the recipient's sign-on scan table, and the word "URGENT" is printed directly under the subject title.

Special delivery options can be entered singly or in combination with one another, and they can be entered following any To: or Send? prompt. They are always enclosed in parentheses. If more than one option is used, each must be separated by a comma. To enter an option at the "To:" prompt, type the name of the recipient followed, in parentheses, by the desired delivery option.

> To: RWILLIAMS(URG,REG),JSMITH(PRI)

To enter an option at the Send prompt, type "N" at the prompt to return to Command level, and then type "SEND" followed by the delivery option.

> Command? SEND (URG)

You also have the option of saving a created message as a workspace under a workspace name you provide. A workspace is a part of each user's catalog and is located in the storage area. Text created in the workspace can be modified, sent, or stored. The workspace differs from

Telemail

files in that files cannot be sent or modified by editing. Use the workspace to hold messages that you expect to change or send, and use files for messages you expect to store only. Messages or text can be edited, purged, unpurged, and saved in files.

To save a message in the workspace you must first create a file in which to place it. The rules for creating workspace files are the same as those for other files. (See the "File Command" section of this chapter.) To create a workspace file, type "N" at the Send prompt, and at the Command prompt, type "SAVE AS" followed by the name of the file you want to create. Saving and filing messages are similar in that they are both methods of storing. They differ in that filing associates any number of received messages or saved workspaces with one filename; while saving associates a unique name with each workspace. A variety of options are available for manipulating saved files. These are discussed in the "Filing and Retrieving A Message" chapter in the Telemail User's Guide.

Several deferred-delivery or repeat-delivery options may be used with the Send command. Notice that the options are always enclosed in parentheses.

Send (After date time). Used to send a message as soon as possible after the date and time specified in the command.

Send (Every time period, xx Times). To send repeat messages every hour, day, week, month or year for a specified number of times.

Send (Every time period Before date time). To send a repeat message every hour, day, week, month or year before a specified date and time.

Send (On date time). To send a message on a specified date at a specified time.

Send (Between date/time And date/time). To request that a message be sent between the specified times and dates.

There may be instances when you want to cancel a deferred-delivery or repeat-delivery message. To do this, type "CANCEL" at the Send prompt. You are asked for the system message number of the message to be cancelled. Remember, this is the number in the upper right corner of every received message, and in the lower right corner of messages you send. You will receive a notification of the cancellation.

UPLOADING FILES TO TELEMAIL

The most cost effective and efficient way to send messages is to prepare them using your own computer and word processing software and then upload them at your convenience to Telemail. Messages sent in this way must be prepared in a specific format so Telemail can read the envelope information correctly. For example, Telemail always prompts for CC: information. If your prepared message includes a CC: enter it in the usual way. However, if it does not, you must press Return at the point a CC: prompt would normally appear. This tells Telemail to leave this space blank and advance to the next field. Be certain to type a period and then press Return at the end of the message, just as you do when preparing messages online.

To transmit a message, logon to Telemail and type your username followed by an exclamation point ("!"). This suppresses the welcome banner and notices. At the Command prompt type "Compose Telemail.Batch" and press Return. Then take whatever action your software requires to transmit messages. When the transmission is complete the Command prompt appears. Type "STATUS (STA)" followed by Return.

Telemail provides a status log that enables you to verify that the message was sent without error. Messages containing errors are placed in a temporary workspace within your catalog where they remain for up to 24 hours. This gives you an opportunity to review them and take appropriate action. Refer to "Creating Messages Offline" in your User's Guide for more information.

USING MAILING LISTS

Mailing lists are prepared by the user, but maintained by the organization's ADMIN. The Compose List command is used to submit information to the ADMIN. When you use the Compose List command, you must enter the abbreviation for a hierarchial level, which shows who has access to the list. The abbreviations are:

SYS	Anyone in your system
ORG	Anyone in your organization
DIV	Anyone in your division
SUBDIV	Anyone in your subdivision
SEC	Anyone in your section
SUBSEC	Anyone in your subsection

Telemail

To create a new list, type "COMPOSE LIST" at the Command prompt. You will be asked if you want to create a new list, add a member, or delete a member. Type "CREATE" and press Return. At the Command prompt, type the name you want to give the list and press Return. When prompted for who has access to the list, type one of the hierarchial levels shown above and press Return. You are now ready to enter the usernames of the members to be included on the list.

At the Member prompt, type the first name and press Return. You will continue to receive the Member prompt until you press Return at a prompt without entering a name.

>Please enter the list name: BOWLERS
>Who can access the list: DIV
>Member: JJONES
>Member: PSMITH
>Member: RROBERTS
>Member: (Return)

To display a list of names, type "MEMBERS OF" followed by the name you gave the list.

>Command? MEMBERS OF BOWLERS

Members names can be easily added or removed from a list. Refer to "Composing A List" in the User's Guide.

SCANNING MAIL

"Scan" (S or SCA) is an important command in Telemail and may be used at any Command prompt. It is used to obtain a chronological table of all unread messages in your mailbox and is the primary command used to retrieve and read these messages. Scan can also be used to locate, display, and act upon any file or workspace in your catalog. Each message received is assigned a unique scan number, starting with one, and is listed in the scan table in the order of receipt, the oldest first. An exception is a message tagged "Urgent." Urgent messages are placed at the top of the list. Scan can be typed at any Command prompt.

The list displays the scan number of the messages, the date and time posted, the username of the person sending them, the subject, and finally, the number of lines they contain. The list is followed by the Command prompt. The system automatically pauses at the end of each 20 lines and prompts for "More." To suppress the More prompt, type "Scan!" at the prompt. The "!" can be used with any scan command.

Scan Options

A number of options are available for use with the Scan command.

Scan (Action). Generates an "Action?" prompt after each item in the scan table. This permits the user to handle each item as it appears on the screen. It can be used with any scan option.

Scan All. Displays a table of information on every message in the catalog. Data is presented in delivery date order beginning with the first unread message.

Scan Before date. Scans messages delivered before a date you specify.

Scan File filename. Scans all messages filed in the named file.

Scan From username. Displays a list of all messages sent by a particular person.

Scan On date. Scans a list of messages delivered to you on a specific date.

Scan Purge. Displays a list of messages purged in the last 24 hours. Purged messages are retained in the system for a period of 24 hours. They cannot be retrieved after that time.

Scan Since date. Displays a list of all messages received on or after the date specified in the command.

Scan Subject "string." Displays a list of all messages containing a specific subject title. Always enter the string (subject) in quotes.

Scan (Summary). Displays a one line summary of messages in your catalog that are identified by one of the scan options discussed above. It can be used with any scan command.

READING MAIL

The "Read" (R or REA) command displays messages that appear in the scan table or in specified files in your catalog. When you logon to

Telemail

Telemail, the welcome screen displays a list of new messages in your mailbox. Let's see how easy it is to read a message.

Type "READ" at the Command prompt. Messages are displayed in the order of their scan numbers unless a specific message is requested by entering a scan number following the Read command. You can also request a series of messages by typing scan numbers separated by a comma, or request a range of scan numbers separated by a hyphen. Any combination of these can be used.

    READ 3
    READ 1,5,7
    READ 9-15
    READ 3,5,7,9-15

Each message has a number in the upper right corner. This is called the System Message Number (SMN) and is used by Telemail to keep track of all messages in the system. There will be occasions when you must use the SMN as part of an action being taken. We'll discuss these later.

Read Options

Several options are available for use with the Read command.

Read All. Displays all read and unread, filed and unfiled messages in your catalog.

Read! Permits you to read all messages without taking an action on them.

Read Before date. Displays messages received before the date specified in the command.

Read scan number. Allows you to choose which messages you want to read.

Read File filename. Allow you to read every message in the specified file.

Read From username. Displays all messages received from a specific Telemail user.

Read On date. To read messages received on a specific date.

Read Purged. Displays every message purged in the last 24 hours.

The Electric Mailbox

Read Since date. Displays every message received since a specific date.

Read Subject "string." Displays every message in your catalog that contains the specified string in the subject field. Be certain to enclose the string in quotes.

Read Unfiled. Displays every message in your catalog that has not been placed in a file.

Read Unread. Displays all messages in your catalog that have not been read.

Telemail also provides an Unread (UNR) command that places the just-read message back in the mailbox where it remains until requested again. However, since the message has been read, storage charges will begin after five days storage. The Unread command can be used with most of the Read options, except Read Purged. A purged message must be unpurged before it can be read.

An "Action?" prompt follows each message. The system is asking what you want to do with the just-read message. You can answer the message, forward it to another user, purge it from your files, or file it in your storage area for later action. If you do not want to take an action on the just-read message, press Return, and the message will be returned to the mailbox. Remember, you can type "Read!" to suppress the Action prompt. To bypass the Action prompt and return to the command level, type "EXIT" at the Action prompt.

Let's take a look at the available options for handling read messages. At the Action prompt, you can enter these commands.

Answer

The Answer (A or Ans) command allows you to immediately respond to the sender and all other recipients of the just-read message. This command can be used at any Action or Command prompt. You are prompted for the text of the message. It is not necessary to enter the name of your correspondent, since this information is taken from the message being answered. To end the message, type a period on a line by itself and press Return. This will generate the Send command. If you are ready to send the message, enter "Y." The command is confirmed and you are returned to the Action prompt. But if you want to take another action on the same message, type the new command at the prompt. If you don't want to take

Telemail

an action or to enter a command, press Return to display the next message.

When you use the Answer command by iteself, your response goes only to the sender of the original message. Anyone who received a courtesy copy (CC:) of the message will not receive the response.

To respond to all recipients of the original message, use the Answer All (A ALL) command. You are also prompted for additional usernames to which you would like to send your response.

Forward

The Forward (F or For) command is used to send a just-read message to another user. The option can be entered from the Command or Action prompt. A message can be forwarded just as it is received, or comments can be added to the beginning of it. To do this, type "FORWARD" at the Action prompt. You will be asked for the name of the user to whom the message is to be sent, courtesy copy names, and a subject title. More than one name can be entered on the same line, but each name must be separated by a comma. You are then prompted for your comments. A period on an empty line followed by pressing Return signifies the end of your comments and generates the Send prompt. Here are the options that can be used with the Forward command.

F All. Forward All. Forwards all messages, including read and unread, filed and unfiled.

F Before date. Forwards a copy of every message in your catalog with a date before the date named in the option.

F File filename. Forward File. Forwards every message currently filed in a file you name.

F From username. Forward From. Sends every message received from the specified Telemail user to a recipient of your choice.

F On date. Forward On. Forwards a copy of every message received on the date named.

F scan number. Forwards a specified message from your mailbox to another user.

The Electric Mailbox

F Since date. Forward Since. Forwards a copy of every message received since the date you specify.

F Subject "string." Forwards a copy of every message in your catalog that contains the specified string of characters in the subject field. Remember to enclose the string in quotes.

Purge

The Purge (Pur) command is used to delete unwanted messages from your catalog. However, be careful when using this command since it erases the message from the system and it will be impossible to retrieve after 24 hours. (See Unpurge.) If you request a message to be purged that resides in several files, Telemail informs you of the number of files the message is in and asks whether you still want to delete it. If you answer "Y," the command is carried out. Otherwise the Purge command is ignored and you are returned to the Action prompt.

Several options are available when using the Purge command.

Pur Before date. Deletes all messages with a date before the date entered following the command.

Pur File filename. Purges the named file. The file and all messages currently residing in it are erased.

Pur From username. Erases all messages received from the username specified in the command.

Pur On date. Causes all messages delivered on the date specified in the command to be erased.

Pur scan number(s). Erases specific messages from the scan list.

Pur Since date. Erases all messages received since the date named in the command.

Pur Subject "string." Erases all messages whose subject contains the word(s) specified in the string. The string must be enclosed in quotes.

Pur Unfile. Erases all unfiled messages in your catalog.

Pur Unread. Erases all messages you have not read.

Telemail

Unpurge

Once a message is purged, it is possible to retrieve it if you take an Unpurge (Unp) action within 24 hours of the time the message was initially erased. Telemail maintains a list of all messages purged within the last 24-hour period. To review the list, type "SCAN PURGED" at the Command prompt. Remember, you cannot take an action on a purged message unless it is Unpurged first. This command can be entered at the Command or Action prompt. The following unpurge options are available.

Unp All. Unpurges all purged messages, including read and unread, filed and unfiled.

Unp Before date. Unpurges messages received before the date specified in the command.

Unp From username. Unpurges messages received from a specified Telemail user.

Unp On date. Unpurges messages delivered on the named date.

Unp scan number(s). Restores specified purged messages.

Unp Since date. Unpurges messages delivered on or after the date specified in the command.

Unp Subject "string." Unpurges messages that contain the specified string. Be certain to enclose the string in quotes.

File

With the File (Fil) command, files can be established within your catalog storage area to enable you to organize messages conveniently for easy retrieval. A single message can be placed in up to six different files. To create a file, type "FILE" at the Action prompt followed by the name you want to give the file. A filename can contain up to 20 alphanumeric characters, but must begin with an alphabetic character. Spaces and special characters are not permitted in filenames, and the only punctuation allowed is the period. "Production" and "Production.Report.2" are valid filenames. Several file options are available.

Fil All In filename. Files all messages including read and unread, filed and unfiled in a single file.

Fil Before date In filename. Files all messages received before the specified date in a file named by you.

Fil filename1 In filename2. Permits you to combine files by entering the contents of one file into another file.

Fil From username In filename. Files all messages from a specified Telemail user in a file you name.

Fil On date In filename. Files all messages delivered on a certain date.

Fil scan number(s) In filename. Permits you to file a message you identify by a scan number in a file named by you.

Fil Since date In filename. Files all messages received on or after the specified date in a file named by you.

Fil Subject "string" In filename. Files all messages with a subject you specify in a file of your choosing. Be certain to enter the string in quotes.

Fil Unfiled In filename. Clears your desk of unfiled messages. For example, you can use it to hold messages until an action is taken on them.

To display a list of all files you have created type the command "DISPLAY FILES." In addition to the above options, valid commands also include "Scan File filename" and "Purge File filename." These options were discussed earlier.

SHORTCUTS

More than one command can be entered on the same line. This is done by separating each command with a semicolon. For example, the following command line allows you to read all your new messages without response, file them in a file named "Letters," and logoff the system.

    READ!;FIL ALL IN LETTERS;BYE

As each command is executed, you are notified that the system is responding to the next command.

    ***Next cmd:READ!

EDITING MAIL

Messages you are currently composing, as well as messages previously created, may be edited using the Telemail editing commands. To edit a just-composed message, type "N" at the Send prompt, and when the Command prompt is displayed, type the desired edit command.

The Edit command is used to create and save text, but can also be used in Compose mode. The commands used in both situations are the same; the method of entry is different. In Edit mode, commands are entered following a Command prompt, while in Compose mode they are entered as dot commands. To learn more about the Edit function, consult the Telemail User's Guide, or type "HELP EDIT" at the Command prompt.

OTHER TELEMAIL SERVICES

INFORM Scripts

Customized standard forms for call reports, price lists, memos, promotional announcements, and other business needs can be created using Telemail's INFORM (INformation FORmatting) script option. Information entry is easy, and automatic validation of responses helps eliminate error. If you are interested in creating you own scripts, contact your ADMIN for a copy of the INFORM script guide.

Direct Delivery Service

Direct delivery service enables Telemail users to send messages not only to other Telemail subscribers, but to more than one million organizations throughout the world. Messages can be delivered to any terminal if it is connected to a domestic or overseas Telex or TWX network or to a Telenet public data network by leased lines. Not all Telemail users are permitted to use this service, so check with your ADMIN to determine whether you have access.

Telex

Although your organization may already have Telex or TWX equipment, the ability to access these services through the Telemail system enhances your productivity and enables you to edit and correct messages before they are sent. You need only key your message once to send the same message to Telex and Telemail users.

Dial-Out Direct Delivery (DDD)

When using DDD, Direct Distance Dialing, long-distance telephone lines are used to "call" a specified modem-equipped, auto answer terminal (usually with a printer) capable of sending or receiving "telephone calls." Incoming calls can be received at anytime, even without an operator in attendance, as long as the terminal is turned on. The terminal must be compatible with the Telemail system. This service is available anywhere in the United States. Telemail dials the number, makes the connection, and delivers the message. The recipient of the message does not have to be a Telemail subscriber.

Net Direct Delivery

The Net Direct Delivery service permits terminals connected to the Telenet network by a Dedicated Access Facility (DAF) to route messages to designated network addresses anywhere in the continental United States. International deliveries can be made to countries such as Germany, Mexico, Switzerland, and the United Kingdom.

Station Direct Delivery

A station can be any delivery device (DDD, Net Delivery, Telex, or TWX) located in a Telemail subscriber organization or not. Since a Telemail station is registered on the Telemail system, it is possible for the user to address a message with a Telemail name rather than the longer more complex address required on the other direct delivery services. Stations must be registered by your ADMIN.

Telemail also provides a way to receive a hardcopy of all incoming messages to a specific station. A second copy is sent to your personal catalog. This option is available for all but Private and Registered messages. The service must be arranged for with your Telemail Administrator.

Telemail makes sending direct delivery messages easy. But there are a few guidelines that must be followed, especially in the "To:" and "CC:" address fields, where information must be entered in a specific way. For all direct delivery messages, the recipient's name can consist of no more than 20 characters. Punctuation, except for commas, is permitted, and the name can be typed in upper- or lower-case letters. Follow the name with the address of the receiving device (DDD, TLX, TWX, or NET), a colon and the recipient's number or letters enclosed in parentheses.

Telemail

An answerback (Ans) can be included if the receiving device supports this function. An answerback is a string of characters by which a telex device is identified. A message is not delivered until the correct answerback is given. The answerback can consist of no more than 20 characters and must be followed by a one-character poll character (CHR:) to determine the receiving terminal's readiness.

TO: CRobins (TWX: 246-810-1214, ANS:ROBNSC,CHR:CTRL-E)

When using the station direct delivery option, determine the name of the station closest to your recipient by typing the "Station Of" directory command. At the "To:" prompt, type the name of the recipient followed, in parentheses, by "STN:" and the station name.

To: RRoberts (STN:CPMCLUB)

Domestic and international telex devices are reached by using the Telex Direct Delivery option. A telex address begins with the recipient's name and is followed by three information fields: area code, telex network address, and answerback (if used). The recipient's name must be entered in full. This is because many organizations use one telex device for all employees.

The area code (AC:) is the first part of the address field and is commonly called the "country code." It identifies the country to which the message is being sent. A complete list of country codes can be obtained by typing "COMPOSE COUNTRY CODE" at the Command prompt.

The Telex network address (TLX) can consist of 3-9 numeric characters. If the address is directed to a Western Union device, it must begin with a zero (0).

An answerback is an optional entry, but an advisable one since it ensures that your message is sent to the appropriate telex equipment.

All the information following the recipient's name must be enclosed in parentheses, with key information separated by commas. The same message can be sent to other Telex or Telemail users or to any other direct delivery device. A return receipt (REC) can be requested if you want confirmation of delivery of your message. Type "REC" enclosed in separate parentheses following the full address.

To: Samual Brown (AC:421,TLX:43212,ANS:CORP) (REC)

The Electric Mailbox

Keep in mind that some telex devices are restricted to 64 characters per line. Also, certain characters are converted or expanded to other characters. For example, "*" is converted to "=" and "$" is expanded to "DRS." Additionally, you may not use "NNNN" or "LLLLL" in either upper- or lowercase.

LOGGING OFF

To logoff the Telemail system, type "BYE" at the Command prompt. Telemail will display session data, including the network address, the time spent on the system, and the number of packets of information sent.

CONCLUSION

GTE Telemail is a full-featured electronic mail and information distribution system that provides effective and efficient communication services throughout the world. It is a reliable means of communication for large business organizations. But it is also feasible for an individual or a small business owner via Telenet's reseller program. Speed, convenience, and flexibility are the key features of this sophisticated electronic mail system.

RCA Mail

RCA Mail
RCA Global Communications, Inc.
201 Centennial Avenue
Piscataway, New Jersey 08854
800-526-3969

RCA Global Communications has been providing communication services for nearly 70 years. It was the first to provide telex services overseas, the first to use satellites to send telex messages for faster and more efficient service to its customers, and the first electronic mail provider to offer receiving and sending of telex messages. In 1984, RCA Globcom introduced RCA Mail that added a new dimension to its customer's communication capabilities. Not only did this new service make it possible for the customer to send messages to other RCA Mail subscribers using their own computer or communicating word processor, but through the store-and-forward feature, to send and receive telex messages without the use of a dedicated telephone line or a telex terminal.

RCA Mail and RCA Telex are two distinctly different services, but are highly compatible and each complements the other. Access to RCA Mail is available from the US and over 40 countries. RCA Telex is available only from the US and possessions. In other countries, that nation's telecommunications administration provides telex which interfaces with RCA Globcom's telex; hence, a worldwide network of over 1.6 million telex terminals. Directly or indirectly, RCA Globcom can provide telex service to any country that has it. They have direct service with about 200 countries.

RCA Globcom's Hotline news service (called FYI outside the US), was introduced in 1983, enabling customers to get news around the clock from anywhere in the world. Hotline provides up-to-the minute business reports on finance, stocks, business, medicine, politics, and weather. Additionally,

The Electric Mailbox

general interest features on topics such as books, horoscopes, movies, and theater are available. Information is immediate and comprehensive.

In 1985, RCA's "Quick 'N EZ Access" was introduced, making it possible to use one local telephone number for access to any of RCA Globcom's telex services. Before the introduction of this service, it was necessary to dial special telephone numbers depending on the baud rate and service being used.

HOW TO SUBSCRIBE

Make application by calling the Customer Care Center at 800-526-3969, or by writing to the nearest RCA Globcom business office.

RATES

RCA Mail and Telex services are available 24 hours a day, 7 days a week. The times given below are for your local time.

Business rates apply Monday through Friday, 7 a.m. to 6 p.m. From the continental US and international points, the rate is $14.00 per hour. WATS access is $18.00 per hour. There is an additional international access charge.

Off-peak rates apply Monday through Friday, 6 p.m to 9 p.m. and Saturdays, Sundays, and major holidays 7 a.m. to 9 p.m. From the continental US, and from international points and for WATS access, the rate is $14.00 per hour. An additional access charge is made for international points.

Night-time rates apply every day, 9 p.m. to 7 a.m. From the continental US the rate is $4.00 per hour; from international points, $14.00 per hour, and for WATS access, $6.00 per hour.

A monthly minimum of $500.00 per subscribing organization is charged after the first three months of service. This fee does not include the subscriber's monthly account charge of $140.00.

There is no charge for the first five days of storage. After that, seven tenths of a cent is charged per day for each unit stored. A unit consists of 1000 characters transmitted in or out of the system on a session basis.

RCA Mail and real-time telex support 300 and 1200 baud. RCA Telextra (store and forward telex) supports 300, 1200, and 2400 baud. Telex terminals are supported at their normal 50 baud rate.

ACCESSING THE NETWORK

RCA Mail is accessed through GTE Telenet or through WATS service. See the appendices for a list of telephone access numbers.

CONNECTION PROCEDURES

To connect to Telenet, dial your local Telenet telephone number and wait for the carrier tone. Press Return twice. The "Terminal=" prompt will be displayed. Type the terminal ID code given to you by your RCA Mail Administrator, and press Return. At the "@" prompt, type "MAIL." Your connection is complete and you can proceed to logon.

LOGON PROCEDURES

When you are connected to RCA Mail you will be asked to enter your user name. Type the user name and press Return. Next, you will be asked for your password. Type your password and press Return.

   User Name? P.ROSENBURG (Return)
   Password? XXXXXXX (Return)

There may be more than one person on the system with the same user name as yours. So you may be requested to identify yourself further by entering your organization's RCA Mail name. To bypass the "organization" prompt, append your organization name to your user name with a separating slash ("/"). You will notice that you cannot see your password, and neither can anyone else. This is a security device that prevents unauthorized people from accessing your personal mailbox.

   User Name?V.LOYD/DSS.CORP

Once you have correctly entered your user name and password, the welcome screen will be displayed, and if messages have been posted to a monitored bulletin board, you receive a message to check those boards. This is followed by a scan list of all new messages received since you last logged on. The list is followed by the "Command?" prompt.

USING RCA MAIL

RCA Mail is a full-featured mail system that permits users to send messages to and receive messages from other RCA users, post messages on bulletin boards, store messages in files for easy reference, and compose, forward, and answer telex messages. RCA Mail uses the same mail sys-

The Electric Mailbox

tem as GTE Telenet's Telemail. The major difference between the two systems is the way user names are entered in the "To:" and "CC:" fields. In Telemail, the name is entered without punctuation. In RCA, however, the user name is always entered with a period (".") following the first initial and before the last name ("R.SMITH" for example).

The user's guides are virtually the same, although RCA has rearranged some commands and added instructions for composing, forwarding, and answering international and domestic telex messages. To use the RCA Mail system, refer to the Telemail chapter in this book.

OTHER RCA GLOBCOM SERVICES

Sending International Telex Messages Via RCA Mail

International telex messages may be sent all over the world from your computer. Composing an international telex message in RCA Mail is easy. But before you can start, you need to be registered as a telex customer (no extra charge). You also need an area code to identify the country to which the message is to be sent, the telex number of your correspondent, and an answerback (optional) so RCA Mail can verify that the telex message is delivered to the correct telex address.

At the Command prompt type "COMPOSE" and press Return. You are prompted for the envelope information, including the "To:," "CC:," and "Subject:" fields. RCA provides an attention field after the "To:" prompt for the purpose of entering the name of the recipient of your message. Since many organizations have only one telex number, this information assists in making certain that the intended recipient receives the message. The attention field can consist of up to 20 characters. A period is not required between the first and last names.

     To: Thomas Smith

The recipient's name is followed by a left parenthesis, the country area code (AC:), and a comma. You can request a complete list of 3-digit area codes by typing "COMPOSE COUNTRYCODE" at the Command prompt.

Next, type "TWX," a colon, and the telex number of your correspondent. Follow the telex number with a comma. If the message is to be delivered to a Western Union telex number, the telex number must be preceded by a zero (0).

RCA Mail

The answerback (ANS:) is entered after the telex number and followed by a right parenthesis. Then press Return. If an answerback is not used, place the right parenthesis after the last digit of the telex number. Although the answerback is an optional entry, its use is encouraged to ensure correct delivery. The address line is now complete and will look like this:

To:Thomas Smith(AC:819,TLX:123456,ANS:ABTCOMM)

After you press Return at the end of an address entry, you will receive another "To:" prompt. You may enter another address or press Return to generate the CC: prompt. Do not enter a comma following the last address. Just press Return.

If you send to the same group all the time, you may store the addresses as a document by separating each set of addresses by a comma. You may upload your own address list in the same way you upload a document. The difference is that instead of typing "COMPOSE," you would type "COMPOSE BATCH." This works with telex numbers, RCA Mail addresses, or a combination of the two.

The "CC:" prompt is displayed next. Courtesy copies can be sent to other telex users or RCA Mail subscribers. Each telex address must be entered on a separate line, but names of RCA Mail recipients can be placed on the same line, separated by commas. If the RCA Mail address entries require more than one line, end each address, except the last one, with a comma to receive a new "CC:" prompt. Press Return following the last entry to display the "Subject:" prompt. Type a short subject line and press Return. You will now see the "Text:" prompt. You are ready to enter the text of your message.

A telex message can consist of as many lines as you want, but should contain no more than 69 characters per line, including spaces. This line width is not critical in store and forward unless you are concerned about the format. In real time, however, it is essential. Press Return after each line of text.

It is wise to enter your telex number in the last line of text so your correspondent will know where to send a return message. An entry such as "Please respond to 123456 ATB UR" would be appropriate. Type a period (".") and press Return on the line following the last line of text to signify the end of the message. This generates the "Send?" prompt.

The Electric Mailbox

A "Y" response to the Send prompt is confirmed by a system's entry of the date and time the telex was accepted and the message number assigned to it. You are then returned to Command level.

    Command?COMPOSE
    To:Johnny Brown(AC:303,TLX:123456,ANS:BRNPALACE)
    Subject:Coming Home

    Text:
    Sailing tomorrow on the Titanic.
    Assuming good weather, I'll see you in New York
    on the 17th.

    Love,
    Molly
    .

    Send?Y

    Msg posted Apr 12 13:47
    Msg:ALDY-9876-5432

Sending Domestic Telex Messages Via RCA Mail

Domestic telex messages are prepared in the same way as international messages, with the exception that they do not require an area code. Type the telex number, and answerback, if used, in the "To:" field, complete the "CC:" and "Subject:" fields, then type your message. If the message is to be delivered to a Western Union telex number, the telex number must be preceded with a zero (0).

    TO:Robert Jones(TLX:0246810,ANS:FOOD)

Telex messages can be forwarded and answered just like any other message created in the RCA Mail compose mode. To answer a telex message, you must send one. There is no automatic answer capability for an incoming telex message.

RCA Telex Services

In addition to RCA Mail, RCA offers a full range of telex services to its customers. In fact, RCA is perhaps best known for these computer-originated telex services. We will discuss each service briefly. For complete details, read RCA's "A Guide To RCA Telex Services and How To Use Them."

RCA Mail

The terminal used to send telex messages may be a computer, communicating word processor, teleprinter or one of many communications terminals available today. In order for your terminal to serve as a telex terminal, it must be registered with RCA Globcom. To register, call your local RCA sales representative or the Customer Care Hotline. You must also have a modem. If you plan only to send telex messages, an originate only modem may be used. But if you want to receive messages in real time, the modem must be capable of automatically answering the telephone 24 hours a day and be connected to a dedicated telephone line. And finally, your signaling protocols must be compatible with RCA Globcom's.

Sending Telextra Store and Forward Messages

Store and forward messages are sent from your computer to the RCA computer where they are stored for a short period and then forwarded to your correspondent. To connect to Telextra store and forward through DDD (Direct Distance Dialing), dial RCA using the DDD number supplied by RCA, or call the Customer Care Center to obtain a number. Your connection is complete when the words "RCA TELEXTRA" appear on your screen followed by a request for your identification, "YR ID PLS?" You may now proceed to enter your ID and message.

To connect to Telextra store and forward from an RCA dedicated terminal, press the Start button on the terminal or follow the instructions in your terminal manual. Once the connection is made, the RCA "GA" (go ahead) prompt appears. Type "134+" to tell the RCA computer that you want to enter the Telextra store and forward mode. You are informed that you are in Telextra store and forward, and your telex number and answerback are displayed followed by the reference line and GA prompt.

Using RCA Real-Time

Real-time telex is communication between two terminals located anywhere in the world. This form of terminal-to-terminal communication is called domestic or international telex. A dedicated telephone is required since the terminal must always be ready to receive a call.

To connect to RCA Globcom's real-time through DDD, dial the appropriate telephone access number. You know you are connected when you see the RCA "GA" prompt. You may now proceed with your session. To send a real-time message from a private tie line terminal, begin by pressing the Start button on the terminal or follow the directions in the manual supplied with your terminal.

Special Features

RCA provides a number of options to make your communications experience more enjoyable and cost-effective by shortening the period of time online. Additionally, several billing options are available, making it possible to track usage of the RCA service.

Unicode/Duocode Selections. Used to dial frequently called overseas telex numbers by using a one- or two-digit code in place of the entire telex number. These codes may be used in both the real-time and RCA Telextra store and forward modes. When you enter the appropriate code, the number is looked up by the RCA Telex exchange and the call is completed. To use the codes, simply enter the code number in place of the telex number.

Departmental and Segmented Billings. Billing charges can be assigned on a per telex basis to different departments within your organization by the addition of a special tag entered following the telex number of your correspondent. To use this feature, type the telex number of your correspondent, a period, and the tag name assigned to your department.

Segmented Billing. The segmented billing option permits charges to be billed to different departments within the same message, for example marketing and management. To use this feature, a code 64 (segmented billing code) must precede the telex number.

Customer Reference Code (CRC). A private reference code of your choice may be added to the telex number of the correspondent. The number enables you to further identify users or, if you wish, to classify messages by type.

Multi-Batch Messages. Batched messages allow you to send more than one telex message during the same session. Multi-batched messages can only be used in the Telextra store and forward mode.

Store and Forward Using Answerback. The answerback verification option is used to ensure that the telex message you enter is delivered to the right terminal. RCA verifies the answerback as it is entered and does not deliver the message unless there is a 100% match.

Multi-Address and "ATTN:to" Line. Telextra store and forward permits you to send the same message to many different recipients by using the multi-address option. In addition, you can add an "ATTN:to" line following each telex number to customize the message.

RCA Mail

Alternate Selection. The alternate selection option is available for RCA Telextra store and forward messages. It permits you to select an alternate recipient for a telex message. This option is helpful in those situations where the primary recipient cannot be reached and it is imperative that the message be delivered to someone who can take action on it.

Select-A-Time. On a per message basis, RCA Telextra provides a way to indicate how long RCA should attempt to deliver a telex message. Six attempts are made during the first hour, and then once an hour for the next twenty-three hours.

These special features can be used in combination with one another in a single message. For example, you can combine Duocode, CRC, send multi-address, multi-batch, ATTN:to, and an alternate selection with segmented billing.

Sending Overseas Telegrams

In addition to telex messages, RCA offers subscribers a number of other options for sending messages, including overseas telegrams, marine services, and a variety of inbound services.

Real-time DDD or PTL Telegram Service

Real-time telegrams must be sent from a telex terminal unless your terminal is programmed to respond automatically to a CTRL E. To send a telegram in real time, access RCA in the way discussed earlier.

Sending RCA Telextra Store and Forward Telegrams

Telextra overseas telegrams are sent from your terminal to a terminal of the telecommunications administration in the overseas country, and they may be sent to individuals who do not have telex terminals. When the telegram is received by the overseas telecommunications administration, it is delivered by telephone, messenger, or mail, depending on the country to which it is sent.

Sending Telexgram

Telexgram is available in Telextra store and forward and enables RCA to convert a message into a telegram if it cannot be delivered by telex. You have the option of selecting the time at which the telex becomes a telegram by using the Select-A-Time option discussed earlier. The message is

The Electric Mailbox

automatically converted to a telegram at 11 p.m. Eastern Time if it has not been delivered by telex.

Inbound Services

Inbound messages are those directed to you from another individual. These messages can be in response to a message sent by you or a new message. A variety of delivery services are provided by RCA Globcom to assure you that all messages are delivered, even if your terminal is busy. However, you must be registered for these services by your RCA representative. The following delivery options are available.

ThruData. ThruData enables RCA to accept your incoming messages and deliver them to you at a specified time, or attempt to deliver them a specified number of times in a given number of hours. When you register for this service, you must indicate the number of times RCA should attempt to deliver the message to you or the number of hours (up to 24) they should attempt to deliver. This option cannot be used with DataBank.

Alternate Delivery. RCA Globcom provides three forms of alternate delivery. You may have your messages automatically delivered to another telex number when it is received by RCA; have it delivered to an alternate address only if your terminal is busy; or have it delivered at the end of either a time period or a series of attempted deliveries as described in ThruData.

DataBank. Use this option to have your incoming messages stored in the RCA computers where they remain until you pick them up at your convenience. The DataBank must be cleared at least once a day. There is a $15.00 per month charge for this service.

Using Marine Services

RCA Globcom's marine services provide a vital communications link with ships at sea as well as offshore remote stations, such as drilling platforms. Nothing is too small for RCA to handle, from an important shipping or cargo message to a simple greeting sent a friend on a cruise. Marine telegrams can also be sent through RCA to an overseas radio station. Consult your reference guide for full details on how to use this service.

LOGOFF PROCEDURES

When you are ready to logoff, type "BYE" at the Command prompt. RCA will confirm the completion of your session.

CONCLUSION

RCA Globcom provides direct access to any other subscriber on the system and, thanks to RCA Telextra, access to over one and a half million telex subscribers around the world. RCA Globcom is compatible with most terminals and can be easily linked to your present communications systems. The system is particularly well-suited for companies that have need to communicate with both domestic and foreign offices on regular basis.

OnTyme

OnTyme
McDonnell Douglas
Applied Communications Systems Company
20705 Valley Green Drive
Cupertino, CA 95014
800-435-8880

OnTyme was introduced in 1980 by Tymshare. A second version, OnTyme II, was later developed, and at one time Tymshare supported two electronic mail systems. Today, the newer version, OnTyme II, is the only email system in effect. In 1983, McDonnell Douglas purchased Tymshare, making the OnTyme system a part of McDonnell's Applied Communications Systems Company.

OnTyme is a full-featured and comprehensive electronic mail system that combines basic electronic mail services with a wide range of advanced communication capabilities. Through the International Electronic Mail Service (IEMS), subscribers can send Western Union Telex I, Telex II, and International Telex messages in addition to Telegrams, Cablegrams, and Mailgrams. These services are provided under a separate contract.

Each company subscribing to OnTyme appoints an individual to act as their OnTyme Account Supervisor. This individual serves as liasion between their organization and OnTyme and is available to the company's employees to assist with problems encountered with the system.

HOW TO SUBSCRIBE

Application for service is made by contacting your local McDonnell Applied Communication System office, or by calling 800-435-8880. Additionally, you may write to the OnTyme address listed above.

RATES

OnTyme is available 24 hours a day, 365 days a year, except for normal maintenance. Maintenance usually occurs in the early morning hours, and a banner always announces when the system will be down. The basic OnTyme service charge from domestic US and Alaska is $3.00 per hour, and $8.00 per hour from Hawaii, Puerto Rico, and Mexico. The WATS service charge is $23.00 per hour. In addition to these fees, a twenty-five cent per 1,000 character input/output fee is also charged. There is a two minute minimum time charge per session.

OnTyme charges a $200.00 monthly subscription fee, and a minimum monthly account charge of $500.00. Discounts are available on accounts totaling more than $5,000.00 per month. If the actual charges on the invoice, including the monthly subscription fee, total less than $500.00, the difference is charged to the customer. File storage is charged at the rate of one cent per 1,000 characters per day. OnTyme supports 300 and 1200 baud.

ACCESSING THE NETWORK

OnTyme is accessed through Tymnet, an advanced public data communication network owned by McDonnell Douglas. Refer to the appendix on Tymnet for a list of access numbers.

CONNECTION PROCEDURE

To connect to OnTyme, dial your local Tymnet access number. When you hear the carrier tone, you know you are connected. You will be prompted for your terminal identifier. The OnTyme Primer provides a complete list of terminal identifiers, but in most cases it is the letter "A." Type "A" and press Return. The screen displays a series of numbers representing the Tymnet access node and the port to which you are connected. These numbers are followed by the "Please login:" prompt.

LOGON PROCEDURES

Before logging on to OnTyme, you must have an OnTyme ID that consists of an account name and an address, a logon code supplied by your Account Supervisor, and a password (called "Keys" by OnTyme).

Your logon usually begins by typing CTRL R followed by your logon code. The "R" is OnTyme's signal to monitor the speed of information flowing to your screen and also enables you to use the CTRL S key to stop the flow

The Electric Mailbox

of information and the CTRL Q key to start it again. Later you will learn how to use CTRL X at logon when uploading files. A banner showing the date and time (Greenwich Mean Time) of your logon is displayed followed by a the system prompt for your ID and password. Type your ID and press Return. At the "Keys?" prompt, type your password and press Return. Your password does not appear on the screen.

    ID? ABTCOM.B/WEST
    KEYS? HOTSTUFF

If your logon entries are correct, the message "Accepted" is displayed, and you are ready to begin your OnTyme session. OnTyme allows you two minutes to complete the logon procedure before automatically disconnecting you. If this should happen, hang up and dial again. You are also given three opportunities to logon with a correct account name and password before being disconnected.

USING ONTYME

OnTyme provides each user with an electronic desk and file. You have a workspace in which to create messages or prepare draft material, an inbox to receive incoming messages, and an outbox in which to place outgoing messages. Another feature of OnTyme is the addition of two more boxes on your desk: an in-old box for incoming messages that have already been read; and an out-old box that contains a list of messages that have been sent. This feature serves as a kind of return receipt that confirms receipt of every message you send. This is automatic, so there is no need to enter a special command to request a return receipt. And finally, you are provided an electronic file drawer where messages and mailing lists can be stored. Messages are retained in the system for 14 days before being deleted.

OnTyme Prompts

OnTyme does not prompt for commands. You must type a colon (:), enter a command, and press Return. Commands are always preceded by a colon. They are OnTyme's signal that you are going to enter a command. Without the colon, OnTyme assumes that you are entering ordinary text and takes no action. You may enter the colon yourself, or at the beginning of each session issue the ":COM" command to instruct OnTyme to automatically supply the colon at the beginning of each line. This is called "Command Mode." When in command mode, you cannot enter messages. You remain in command mode until you type ":TEXT" or logoff the system.

OnTyme

Commands can be typed in upper or lowercase letters or a combination of the two.

Most OnTyme commands can be shortened to two or more letters. In this chapter, the acceptable abbreviations are noted the first time the command is introduced. A complete list of acceptable abbreviations is available in the OnTyme reference manual.

Getting Help

OnTyme provides extensive online help. Simply type ":HELP" and press Return to display a list of commands for which help is available. To receive specific information about a command or its options, type the command name following Help. For example, ":HELP SEND." A good source for obtaining help is the user's guide, called The OnTyme Primer. In addition to the primer, OnTyme offers a reference manual that provides an in-depth look at the system.

To help the user become familiar with the system, OnTyme provides several excellent tutorial files. These can be accessed by typing ":Read *** Tutorial.Basics," ":Read *** Tutorial.Editing," ":Read *** Tutorial.Specialfunction," or ":Read *** Tutorial.Prompting." Each tutorial is written in plain, easy to understand language. If you cannot find an answer to your question, you may contact Customer Service. It is available daily from 6 a.m. to 6 p.m. Pacific Standard Time. The toll-free number to call is 800-435-8880.

Special Control Keys

The functions of the control characters used in OnTyme follow.

CTRL H	Backspace. Deletes the last character typed.
CTRL Q	Deletes a line when editing, or resumes flow of data stopped by CTRL S.
CTRL R	Displays the corrected version of a line when editing, or used to logon to the system.
CTRL S	Halts incoming flow of data.
CTRL W	Deletes the last word typed in a line.
CTRL X	Controls flow of data when uploading files.

Now, let's sit down at our electronic desk and discover how easy it is to use the OnTyme email system. Every time you logon to OnTyme, you are placed immediately in your workspace. The workspace can be likened to the center of your office desk that is the area where you read correspon-

The Electric Mailbox

dence, compose messages, and issue commands to the system. The difference is that your workspace in OnTyme is the area where you control your electronic mail environment. OnTyme compares your workspace to a piece of paper put into a typewriter. Your keyboard, of course, is your typewriter, and your terminal screen a piece of paper that records everything you type at the keyboard. Almost everything you type on your terminal keyboard is put in your workspace. The exceptions are control characters and OnTyme commands.

The Shelf (:SHE) command is an important command to remember. There may be times when you are preparing text in your workspace that you need to place what you are doing aside temporarily to work on something else. This is easy to do using this command. Two Shelf options are available.

:SHE P. Shelf Put. Used to place your current workspace text aside. An alternative to putting text on the shelf is to file it. We'll learn how to file shortly.

:SHE G. Shelf Get. Used to move text from your shelf into your workspace.

SENDING A LETTER

Messages are created in the workspace. There are no commands to tell OnTyme that you want to prepare a message. Simply begin typing the message just as you want it to appear to your correspondent. The exception is that if you used the :COM command to automatically supply the colon, you must type ":TEXT" before you begin your message. Don't use a colon at the beginning of a line of text since OnTyme will assume you want to enter a command.

A line of text can contain no more than 132 characters, and a Return must be pressed at the end of each line. If you don't press Return, the line you are typing wraps around, but after 132 characters, all the following characters will be lost. You receive the message "Line Truncated." Press Return on a line by itself to create a blank line.

When creating messages, simple typographical errors can be corrected using three of the control characters discussed earlier: CTRL H, CTRL W, and CTRL Q. The errors can be corrected only if you are still working on the line where they appear. OnTyme does have an excellent editing

OnTyme

function. You can learn more about it by reviewing the "Editing Messages" chapter of your primer.

Sending Options

You can send a message to a single individual, to several people, or to OnTyme users from different accounts. Once a message is sent, the workspace is empty and you can enter another command or prepare another message.

> Ready to close island deal. Client accepts your
> offer of $24 in trinkets. Please stop by our office
> at your earliest convenience to transfer deed and smoke
> peace pipe.
>
> Sincerely,
> Manhattan Realty Company
>
> :SEND P/MINUIT
> MSG #1626 SENT AT 14 AUG 26 14:25:21

:S address. SEND. To send a message to a single individual, complete your message as described above. Type ":SEND," a space, the address (username) of your recipient and press Return. OnTyme displays the assigned message number and the date and time it was sent. To send the message to more than one person, type ":SEND," a space, and the address of each of your correspondents. Separate each address with a space. You are informed of the message number, and the date and time it was sent. When multiple addresses are sent, each copy is assigned the same message number.

> :SEND R/ROBERTS C/DAVIS F/FRED
> MSG # 12468 SENT AT 12 MAY 85 06:12:54

:S CC address. SEND Courtesy Copy. This option allows courtesy copies to be forwarded to each named individual, and since the name of each person receiving a copy is listed at the top of the message, each person knows that the other received a copy.

> :SEND CC A/BALDWIN D/YOUNG D/BARNES
> MSG #16344 SENT AT 06 JUN 85 17:14:33

The Electric Mailbox

:S accountname.address. To send messages to another accountname, type ":SEND," a space, the accountname, a period, and the address, and press Return.

:SEND ABTCOM.C/DAVIS

You may also send the same message to persons in your own account. Just be sure to include the account name for persons outside your account.

:SEND B/BASHFORD XYZCO.L/LANGFORD N/JONES

Messages you send remain in the system for 14 days. After that, they are deleted by OnTyme. Messages may, however, be stored in a file for permanent retention.

UPLOADING FILES TO ONTYME

Rather than creating text online, you might find it more convenient to prepare your message using your own word processor and then upload it to OnTyme. There are two commands you must know to accomplish the transfer, "Load" and "Transparency." And remember, when you upload files to OnTyme, you must logon to the system with a CTRL X before your logon code rather than a CTRL R. This code requests the network to control the flow of input to prevent loss of data.

Load (:LOA)

When you logon to OnTyme, :Load is set to "off," which enables the system to check any data you send for line length, provide a line feed whenever a Return is typed, and echo your command back to your terminal. However, when you transmit data such as text files and spreadsheets that contain control characters or binary files, you may need to change the transfer environment. OnTyme provides the :Load On (:LOA ON) command to do this. This command enables OnTyme to ignore line length, to translate a carriage return/line feed sequence to a carriage return, and to prevent commands from being echoed back to your terminal.

Transparency (:TRA)

The Transparency command is used with the Load On command to place you in a special OnTyme environment. When you type this command, you will not see any of the characters you type displayed on your screen. This is because it prevents characters from being echoed. You turn Transpar-

OnTyme

ency Off by pressing the Break key on your keyboard. If you do not have a Break key, use the :Set Break command to define a break character. For example, to define the plus ("+") key as the Break key, type ":Set Break +" and press Return.

To upload text to OnTyme, type ":LOAD ON" and press Return Then type ":TRANSPARENCY" and press Return again. Now take whatever action your software requires to send the file. When you receive the "Upload Complete" message, press the Break key to return to Command mode. This causes Transparency to be turned off and the OnTyme colon to be displayed. Type ":LOAD OFF," and at the next Colon prompt type "TEXT." You can now proceed with your OnTyme session.

USING MAILING LISTS

Nothing can be more time-consuming and tiring than sitting in front of your keyboard typing long lists of names of those with whom you correspond frequently. OnTyme saves you this drugery by providing a way to prepare mailing lists ahead of time. You can give the list a name and file it for later use. Create the list in your workspace, but be certain that the workspace is empty before you begin.

To clear the workspace, type ":ERASE" and press Return. Type your list, entering each correspondent's user name on a separate line followed by pressing Return. If the user name is in a different account, be sure to type the account name before typing the user name. When the list is complete, name it and use the :FILE command to file it.

```
:ERASE

V/LOYD
D/YOUNG
ABTCOMM.P/ROSENBURG
T/ALVEREZ
M/JOSEFFER

:FILE * PROMOTIONS
TEXT FILED AWAY AS: * PROMOTIONS
```

Using the mail list is easy. Simply prepare your message, and when you are ready to send it, type ":SEND," a space, an asterisk ("*"), another space, the filename of the mail list, and press Return. The action is confirmed by OnTyme.

The Electric Mailbox

    :SEND * PROMOTIONS

Notice that the filename is preceded by an asterisk (*). OnTyme calls this the "directory indicator." In OnTyme, there are three types of files: private, shared, and public. One asterisk indicates a "private" file; two asterisk identify a "shared" file; and three asterisk designates a "public" file. Since your mail list is "private," one asterisk is used.

READING MAIL

Before new mail can be read, the In-Box must be scanned. To do this, type ":IN-BOX." This command generates a list of all messages in the box. The list shows the message number (a special number assigned by OnTyme to each messages), the ID of the sender, the date and time the message was sent, and the number of lines or characters in the message. Two options can be used with the :IN command.

:IN O. In Old. Displays a list of all messages that have already been read. After a message is read the information about it is transferred from your In list to your :In Old list, and transfers it from the sender's Out list to his Out Old list. We will discuss Out lists shortly.

:IN O SH. In Old Short. This command is just like :In Old except that the list of messages does not show the date and time the message was sent or the sequence number.

After checking the inbox, you will probably want to read one or more of the messages. Messages scroll up the screen to make room for additional text, but you can stop the scroll by pressing CTRL S. This gives you a chance to read the text before it disappears. To begin the scroll again, press CTRL Q. Four options are available for reading the contents of the inbox.

:REA. Read. Displays the first (oldest) message on your In list. Subsequent entries of this command display all other messages in the order they are listed in the table. After you read the message, it is placed in your In Old box and is moved from the sender's Out box to his Out Old box.

:REA A. Read All. Displays all messages listed in your In list in the order in which they were received.

:REA messagenumber. After reviewing the In list, you can choose to read a specific message. Type ":READ" followed by the number of the

OnTyme

message you want to read. Remember, the message number is in the far left column of the list.

:READ E83746

:REA LAST. Redisplays the last message you read. If at some later time you want to reread a message, you must first check your In Old list to obtain the message number and then type ":READ" followed by the message number.

The Break key may be used to interrupt a reading session. If Break is pressed when you are in the middle of the message, the message remains in your In list until you request to read it again. If you finish reading the message before the Break key is struck, the message is placed in your In Old box.

OnTyme has a special way of informing you when a message has been read by the recipient. Once a message is sent, it is placed in your Out box where it remains until the recipient reads it, or it is deleted from the system after 14 days. So by checking your Out box, you can determine whether your correspondent has read the message. To obtain a list of all messages in your Out box, type ":OUT." The list displays the message number, the name of the recipient, and the date and time the message was sent. If the same message was sent to two or more individuals, the message number and names continue to appear on the Out list until all recipients have read it. An asterisk (*) is placed next to the names of those who have read it.

When a message is read by the recipient, it is removed from the Out box and placed in the Out Old box. By typing ":OUT OLD," you can quickly review the list to determine who has read messages sent by you. The list displays the message number, name of the recipient, date and time sent, and the date and time it was read by the recipient.

There may be times when you want to cancel a message that has already been sent but not read by the recipient. First you need to check the Out list to see whether the message is still listed and to obtain the message number. If it is still listed, you know that it has not been read by the recipient. The Cancel command is used to delete these messages from the recipient's In box. The following options can be used with the Cancel command.

:CAN messagenumber. Type ":CANCEL" followed by the message number. You are informed by OnTyme that your command has been accepted.

The Electric Mailbox

:CANCEL E12356

:CAN O messagenumber. Cancel Old. It is also possible to cancel a sent message that has already been read by the recipient. The message is removed from your Out Old list as well as the In Old list of the recipient. Should the recipient attempt to retrieve the message, it will be missing from his box. When the message is cancelled, OnTyme informs you that the recipient has already read it.

CANCEL E90126
E90126 ALREADY READ MESSAGE

FILES AND FOLDERS

Management of information is easy with OnTyme because of its advanced filing system. There are three types of files on the system: private, shared, and public.

Private files are those created by you for your own personal use. They are used to store messages received from others or that you create and send to others, text you create in your workspace, and mailing lists of the names of persons with whom you correspond frequently. When requesting private files, the filename must always be preceded by an asterisk (*).

Shared files are files to which all users in your company's OnTyme account have access. These files should be checked frequently since they may contain important notices to all staff or announcements of interest. When asking to read a shared file, always precede the filename with two asterisks (**). To create, maintain, replace, and delete these files, you must have proper authorization from your Account Supervisor.

Public files are those to which any user of the OnTyme system has access. These files are maintained by the system and are used to disseminate information of interest to all OnTyme subscribers. Public files are designated by three asterisks (***) in front of the filename.

Creating A File

You must create a file before placing anything in it. A filename can consist of a drawername and a filename, or just a filename. A drawername is a prefix to a filename and is used to group related files together. For example, you might have a drawername "STAT1986" in which you file separate statistical files labeled January, February, March, etc. A Drawername can

OnTyme

contain up to 8 characters and is always separated from the filename by a period ("."). For example, "STAT1986.DECEMBER."

A filename is a unique name that identifies a particular file within a drawer. The filename is constructed of any word or series of characters that has no more than 16 characters. You are not permitted to use any word or series of characters that are found in OnTyme commands. For example, you can not use the file name "Production File" since "file" is a valid OnTyme command. Some examples of acceptable filenames include Production Repts, Management Stats, and Public Relations.

Use the command ":FILE * drawername.filename" or ":FILE * filename" to prepare the file. Remember to type an asterisk between the command and the name of the drawername or filename. Note also that there is a space on each side of the asterisk. OnTyme confirms the creation of a file.

    :FILE * STAT1986.February
    TEXT FILED AWAY AS: *STAT1986.FEBRUARY

Filing A Message

Earlier we learned that a message can remain in your In or Out list for up to 14 days before it is deleted from the system. Let's suppose that you decide to place a message currently in your In Old box into a file. Before you can file the message, you must obtain the message number. Use the ":Get messagenumber" command to bring the message into your workspace so it can be filed. The Get command is confirmed by the message "Accepted" displayed on your screen. You can now file the message using the File command.

    :GET E123456
    ACCEPTED

    :FILE * PRODUCTION FIG
    TEXT FILED AWAY AS: * PRODUCTION FIG

A number of options are available for working with files. The following is a brief review of some of these options. To obtain more information, refer to the Ontyme Reference Manual.

:LIS *. List. Displays a list of your private files.

:REA filename. Read. To read a file named by you.

:READ * MEETINGS

The Read command does not transfer a copy of the file into your workspace, it only displays its contents. If you want to work on the file or take some other action on it, you must first use the Get command to transfer the file to your workspace and then enter the Type command (:TY) to display the contents. The file can now be edited or, if you want, sent to another user.

:CLE * filename. Clear file. Used to delete a file. Before removing a file from your file drawer, it is always a good idea to check the contents to be sure that the file can be deleted. This is important because once a file is removed it cannot be retrieved.

OTHER ONTYME SERVICES

Through the International Electronic Mail Service (IEMS), subscribers can send Western Union Telex I, Telex II, and International Telex messages, in addition to Telegrams, Cablegrams, and Mailgrams. These services are provided under a separate contract. Information can be obtained by accessing online public files. Type ":LIST ***" to obtain a complete list of all public files, and then type ":READ" followed by the name of the file containing information about the service in which you are interested.

LOGOFF PROCEDURES

Several commands are available to logoff OnTyme.

:Exit and **:Quit** may be used interchangeably. Use either command to end your OnTyme session and cancel the connection. You are warned if you have incoming messages or if there is text in your workspace or on your shelf, and then asked if you still want to logoff. If you answer "Y," the session is terminated. Incoming messages are saved, but text in your workspace or on your shelf is erased.

The **:Logout** command terminates both your session and connection, and no warning messages are displayed. Any incoming messages are saved, but if you have text in your workspace or on your shelf, it is erased.

Use the **:Signon** command when you want to end your current session but remain connected to OnTyme. You might use this command when you want to logon under a different ID. As with the :Exit and :Quit commands, you are warned if you have incoming messages or if you have text in your workspace or on the shelf.

OnTyme

OnTyme automatically drops you from the system in any situation where you are logged on, but fail to use the system for a period of time. The actual time is predetermined for your account, but is usually 15 minutes. If this should happen to you, your incoming messages and text in your workspace or on your shelf are saved for your next online session.

CONCLUSION

OnTyme is a comprehensive electronic messaging system suitable for large organization. It provides advanced communications capabilities and a broad range of business services. OnTyme is easy to learn and use, very flexible, and highly reliable. It is user-friendly and designed for even the most discriminating user.

GE Quik-Comm

GE Quik-Comm
General Electric Information Services
401 North Washington Street
Rockville, MD 20850
800-638-9636
301-340-4000

General Electric Information Services Company commercially introduced the Quik-Comm System in April, 1981. The system was originally designed for use by GE Information Services in an effort to manage its worldwide operations, and has been used successfully within the company for more than 10 years. The service is designed to complement other network-based applications for corporate users of the company's MARK III teleprocessing network, which is accessible from more than 750 cities in over 30 countries on five continents. The network is available 24 hours a day, 365 days a year by a local telephone call from more than 90 percent of the free world's business telephones.

In addition to email services, the Quik-Comm system also offers a bulletin board capability, a paper mail product called Quik-Gram, telex access, and a software connection between Quik-Comm and PROFS (the Professional Office System from IBM).

A software package, Personal Computer Mailbox, is offered to users of the IBM PC or IBM PC/XT. The software is not required in order to use Quik-Comm, but it allows the users to deal with electronic mail in terms familiar to them. Quik-Comm may be used alone or in combination with other GE Information Services products.

HOW TO SUBSCRIBE

Subscribing to Quik-Comm is as easy as picking up your phone, calling one of the numbers listed above, or the telephone number of a GE Information Services sales office in your city, and asking for more information on subscribing to Quik-Comm for your business.

RATES

Quik-Comm is a transaction-priced system. This means that the originator of the message is charged a fixed price for each type of message transaction depending on the size of the message and the number of recipients to whom it is sent. There are four message classifications: Note (1-300 characters); Memo (300-1500 characters); Document/1st page (1501-3000 characters); and Additional Pages charged in increments of 3000 characters or portion thereof.

Rates for Notes include: file input, 20 cents; offline creation, 35 cents; online creation, 65 cents; additional copies/receipts, 20 cents; and intercountry premium, 5 cents. Rates for Memos include: file input, 40 cents; offline creation, 70 cents; online creation, $1.10; additional copies/receipts, 40 cents; and intercountry premium, 15 cents. Rates for Document/1st page include file input, 55 cents; offline creation, $1.00; online creation, $1.50; additional copies/receipts, 60 cents; and intercountry premium, 25 cents. Additional Page charges include: file input, 45 cents; offline creation, 80 cents; online creation, $1.30; additional copies/receipts, 50 cents; and intercountry premium, 25 cents.

There is a $20.00 per hour communications surcharge for WATS access. There is no monthly minimum charge.

ACCESSING THE NETWORK

Access to the Quik-Comm System is via a telephone call to a local GE Information Services access point or public data network, where applicable. Different types of asynchronous devices, word processors, personal computers, and portable terminals can be used to send or receive Quik-Comm messages via the GE Information Service's teleprocessing network. High-speed devices can also access the service, and many corporate clients of GE Information Serivces are now directly interfacing their in-house mainframe equipment with the Quik-Comm system.

The Electric Mailbox

CONNECTION PROCEDURES

To connect to Quik-Comm, dial the number provided by GE Information Services Company and wait for the "Ready" prompt. To make connection, type "/QUIKM***" press Return, and proceed to logon.

If an 800 WATS line is used, the user can refer to GE's access directory (published quarterly) for the appropriate telephone number in his area.

LOGON PROCEDURES

Before you can logon, you must obtain your Quik-Comm address and password from your company's Quik-Comm administrator. The system will automatically prompt you for your address and password. Type your address and press Return. Then type your password and press Return again.

    Address?IGOT (Return)
    Password XXXXXXXXXXXX (Return)

Your password is typed over a blacked out area and will not print on your screen. Your logon is complete. If you have mail waiting, you receive a Queue list itemizing each message in your mailbox. The list is followed by the Command prompt. You may now proceed with your Quik-Comm session.

USING QUICK-COMM

Quik-Comm is a command driven system. Whenever you see "Command?" Quik-Comm is asking what you want to do. You can make your entries in upper- or lower-case, or a combination of the two. And most commands can be abbreviated to the first three letters. In this chapter, abbreviations for each command will be shown in parentheses the first time the command is used. Some commands must be followed by other information such as a date or message number. Press Return after each command entry.

Quik-Comm also uses "star" commands. These commands always begin with an asterisk (*). Some are used for editing purposes, while others have very specialized uses. You will learn about some of these commands in this chapter. For a full description, review your user's manual.

Quik-Comm

Getting Help

For online help for Quik-Comm, just type "HELP." A toll-free number, 800-638-8730, is available for customer service. Tutorials are not offered online. For further information on the Quik-Comm System, "Your Guide to Office Communications" is available from GE Information Services for $75.00, and includes documentation on all of the Quik-Comm family of products.

With the preliminaries out of the way, let's settle down and learn how to use Quik-Comm. Let's start by learning how to send a simple letter.

SENDING A LETTER

Sending a Quik-Comm letter is easy. Just type "ENTER" (ENT) and press Return. You will be prompted for entries in the"To:", "cc:" and "Sub:" fields. At the "To:" prompt, type the Quik-Comm address of your correspondent and press Return. If you want to send a courtesy copy, type the address of the individual you want to receive the copy at the "cc:" prompt. Then, at the "Sub:" prompt, type a subject heading for your letter. The subject may contain up to 29 characters. If you don't want a subject, just press Return.

The "Enter memo text" prompt is displayed next. Each line is automatically numbered for you, and each number is followed by the greater-than (>) symbol. Type your message, ending each line by pressing Return. To create an empty line, just press Return. No more than 156 characters can be typed on each line. Review Chapter 7 of the Quik-Comm guide to learn how to edit and correct messages. When your message is complete, type "*SEND" (*S) on a line by itself and your message will be on its way.

```
Command?ENTER
To:JALD
cc:MSTA
Sub:DEAR JOHN

Enter memo text:
1>Dear Mr. Alden,
2>
3>I think it unconscionable that you asked Miles
4>to deliver your very personal message to me. Why don't
5>you speak for yourself, John?
6>
7>Priscilla
8>*SEND
```

The Electric Mailbox

Before sending your message, you might find it helpful to have it displayed so you can see just how it will look to the recipient. The *List command is used to do this. Each line will be preceded by the line number, but the greater-than symbol will not appear. You will notice that an item number is assigned by the system. Also, a line number is added to the bottom of the listed message. At this point, if you are satisfied with the letter, you may enter the *Send command.

More than one address may be entered in the "To:" field. You can even enter a short comment for each addressee. To enter more than one address, the successive addresses must be separated by a comma, no space, and followed by a Return.

> To:DWES,NWOD,GEOB

If all addresses will not fit on one line, type a comma after the final address on the line and press Return. The system will display the "More:" prompt. Continue typing addresses as above. A nice feature of Quik-Comm is the ability to add more addresses later if you find that you did not include everyone in the original "To:" or "cc:" fields. You may do this at any point in your letter by typing "*ATO" (add To:) or "*ACC" (add cc:) followed by the additional names. By using the *To or *CC command, you can erase the current fields and add new ones. In fact, you may choose to omit entries in these fields when you first begin your letter and add them later. Just be certain to include them before the message is sent.

The subject field may be corrected by using the *SUB command. Type the command any place in your letter followed by the new subject line. None of the star commands issued while typing your letter will appear in the letter received by the recipient.

You may also add a brief comment following any address. To do this, type the address followed immediately by a space and your comment. The comment may contain up to 14 characters. Follow the comment with a comma and then the next address.

> To:PROS Update,DYON Copy, VLYD Office copy

Following the last entry, press Return and proceed with your letter. Never use a comma in the comment field since Quik-Comm would interpret this as the beginning of a new address. There are several star commands that you will find helpful when creating a letter.

Quik-Comm

*N   *Next.* Sends the letter and returns you to the "To:" prompt. This command is useful when you have more than one letter to send.

*X   *Cancel.* Deletes a letter altogether and returns you to the Command prompt.

*W   *Wipe.* Erases the current message and returns you to the "To:" prompt.

*B   *Build.* Takes you out of the Line Number mode and into the Build mode. You will not be prompted with line numbers or the greater-than symbol. The command may be entered at the first line prompt (1>) or at the Command prompt. When using this command, the "Ready For Input:" prompt is displayed instead of the "Enter Memo Text:" prompt.

When you finish the text of the message, press the Break key to display the next line number. At this point, you may make corrections or type the *Send command.

Quik-Comm provides you with a way to determine whether a message sent earlier has been read by the recipient, and if so, on what date. To do this, use the Display (Dis) command. Type "DISPLAY" followed by the item number, and press Return.

```
DISPLAY 1234567
Status of item: 1234567
sent 86/07/14
KWOF listed 86/07/15
JSMI not listed
```

From the display above, you would know that KWOF read the message on July 15, but that JSMI's message remained unread.

READING MAIL

You will remember that when you first logged on to the system, you received a list showing each message currently residing in your mailbox. The list provides a Queue number (the position of the message in the list), a unique Item (message) number assigned by the system, the name of the sender, length of the message, date sent, and the subject. To read these messages, at the Command prompt, type "LIST" and press Return. (Do not confuse this command with the *List command used when creating a message.)

The Electric Mailbox

Messages are displayed in the order of their Queue number unless you specify a specific message following the List command. Requests may be entered singly or in a series. Always separate each Queue number by a space.

    LIST 2 6 10 16

The Item number may be used instead of the Queue number. Just type "List" followed by the Item number.

    LIST 234987

Several options are available for listing messages when you do not have an Item number. To list all messages on or prior to a specific date, type "LIST I" followed by the less-than (<) symbol and the date. The date is expressed as a single number in the format "YYMMDD." For example, to ask to see all messages on or prior to August 7, 1986, you would type "LIST I<860807" and press Return.

To list all messages on or since a specific date, type "LIST I" followed by the greater-than (>) symbol and the date. For example, to ask to see all messages on or since May 26, 1986, you would type "LIST>860526" and press Return. To list all messages from a specific address, type "LIST I=" followed by the address. For example, "LIST I=SDAV."

To look at incoming mail, but not remove it from your mailbox, use the Scan (Sca) command. This command works just like the List command and has the same options, but it does not delete the items from the sign-on mail list.

A copy of a previous message can be inserted into a current message using the *Copy,n (*C,n) command, where "n" is the Item number of the message to be copied. To use this command, prepare your message in the usual way. When you come to the point where you want to insert the copy, at the line number type the command followed by the Item number.

    10>*COPY,123456

The copied message will include the heading and message. You may use the Quik-Comm editing commands to delete unwanted material.

Use the *File,filename (*F,filename) command to insert a file into existing text. This command works just like the *Copy command.

Quik-Comm

    14>*FILE,PRODSTAT

To forward a message without making changes, use the Forward n (FOR n) command, where "n" is the Item number. Enter a space between the Forward command and the Item number and between the Item number and the address. Note that this is not a star command and that a comma is not placed between the command and the Item number. The address of the recipient of the forwarded message follows the command.

    Command? FORWARD 123654 JTRA

SENDING A QUIK-GRAM

Quik-Gram Service is a paper mail product that enables you to communicate with virtually anyone having a US postal address. It is designed to accommodate specific business functions that require expedited delivery of messages to recipients who do not have equipment to receive electronic mail. These messages are transmitted electronically to sites where they are printed on special stationary, then folded and inserted in matching envelopes. They are placed in the US mail for expedited delivery. Business reply envelopes also may be requested.

The cost of an offline created Quik-Gram is $2.25 per message for the first page of up to a maximum of 40 lines, and $1.00 for each additional page of up to 50 lines. There is a maximum limit of five pages for each message. Add 25 cent per address if a business reply envelope is requested. A 50 cent premium is charged for sending a Quik-Gram using the online creation mode.

Quik-Grams, Quik-Comm messages, and Telex messages may all be sent in the same transmission. For further information on how to send a Quik-Gram message, please check the "Quik-Gram User's Guide."

SENDING A TELEX

Quik-Comm Telex Access is an optional feature of the Quik-Comm system that permits Quik-Comm users in the United States to send and receive messages during a Quik-Comm session. It permits users (once validated for the service) to send messages to terminals on the public telex networks while using the formats and procedures that are used in sending messages to Quik-Comm mailboxes. In like fashion, telex users may send messages to Quik-Comm while using the same formats and procedures they use to send messages to terminals on the public telex networks. For

The Electric Mailbox

further information on how to send a telex message on Quik-Comm, please refer to the "Telex Access User's Guide."

GENIE

General Electric Information Services has a new service for home users, called GENIE. Available during non-prime time hours at a rate of only $5.00 per hour, GENIE offers many of the same types of information services found on systems such as CompuServe and The Source. There is even an electronic mail program that allows you to communicate with other GENIE subscribers. There is an $18 sign-up fee payable at the time of registration, and all fees are charged to your credit card.

GENIE users may exchange messages with Quik-Comm users. To send a letter from your GENIE account to a Quik-Comm address, select "GE Mail" from the main menu. At the "To:" prompt, type the recipient's Quik-Comm address, followed by "@GEISCO" and press Return.

      To: PBUNYAN@GEISCO

To send a letter from Quik-Comm to a GENIE user, at the "To:" prompt, type the recipient's GENIE address, followed by "@GENIE" and press Return.

      To: JAPPLESEED@GENIE

Be sure you type the user's address exactly as it appears in their account information. Otherwise your letter may get misdirected.

LOGOFF PROCEDURES

To terminate your Quik-Comm session, type "EXIT" or "STOP" at the Command prompt.

CONCLUSION

General Electric Information Services Company provides messaging for corporate clients regardless of the sender or receiver's hardware or internal system. The system is easy to learn and use. GE is continually developing and adding other mail products to its system that help the user reap the full benefits of office automation. The system offers sophisticated communications and information access to companies needing to send and receive a substantial volume of domestic and international electronic mail messages.

ITT Dialcom

ITT Dialcom, Inc.
1109 Spring Street
Silver Spring, MD, 20910
312-588-1572
800-435-7342

Dialcom, Inc., now ITT Dialcom, Inc., was founded in 1970 as a timesharing company. Although development on its mail system and related products began in 1975, the first public email service was not introduced until 1978. ITT Corporation acquired Dialcom in November 1982, and Dialcom has expanded rapidly since then. Although Dialcom is primarily targeted to big business, it does provide service to closed groups such as associations, which are set up as separate accounts. Those who have used The Source will recognize many of the commands used in the Dialcom mail system, since the Dialcom people originally created the program for SourceMail.

In addition to the mail service, standard Dialcom services include bulletin boards, a full text editor and dictionary, two-way conversation (Chat), and an electronic appointment and scheduling system. Dialcom's XMAIL feature permits users to send Telexes, Telegrams, Speedmail Letters, Mailgrams, Cablegrams, and Lettergrams. Value-added services are extensive and include Dow Jones News/Retrieval, INFOX Database Management System, NETLINK, NEWS-TAB, OAG, and EPUB, an electronic publishing service.

Each organization selects its own System Manager who is responsible for providing support to the organization's users and for serving as liaison between the organization and ITT Dialcom.

The Electric Mailbox

HOW TO SUBSCRIBE

To subscribe to ITT Dialcom, contact a Dialcom representative in your area. For the name of the representative in your area, contact the home office at 301-588-1572. You will be asked to sign a service agreement and to select one of several rate plans. When provided with validated IDs, you are ready to enjoy your ITT Dialcom experience.

RATES

Dialcom provides three pricing plans. Plan One applies unless another plan is selected by the customer with ITT Dialcom's concurrence. The plans vary in such areas as contract terms, registration fees, monthly minimums, and hourly rates. The rates described below are for Plan One.

A registration fee of $100.00 is made to cover the cost of ID setup during the first 30 days of service. After that, setup is charged at an hourly rate of $40.00.

Prime rates apply Monday through Friday, 8:00 a.m. to 5:59 p.m., Eastern Standard Time. The base rate is $14.00 per hour, and does not include network access charges.

Non-prime rates apply Monday through Friday, 6:00 p.m. to 7:59 a.m., Eastern Standard Time. The base rate is $6.50 per hour, and does not include network access charges.

Hourly baud rate premiums are $3.00 for 1200 baud, and $6.00 for 2400 baud. There is no premium for 300 baud.

Monthly storage is charged at the rate of 50 cents per block (2048 characters), and minimum storage per mailbox (user ID) is one block.

There is a monthly minimum of $100.00.

ACCESSING THE NETWORK

Subscribers may access ITT Dialcom through Telenet, Tymnet, Uninet, Datapac, or public 800 WATS.

CONNECTION PROCEDURES

The exact connect procedures depend on the network being used. And before you make connection you will need a "computer systems number."

Dialcom

The system number is represented by the letter "n" in the following connection procedures. When connection is made, the screen will display "PLEASE SIGN ON." At this point, go to the "Logon Procedures" section of this chapter. See the appendices for a list of access numbers.

Uninet. For Uninet access, dial your local Uninet number. When connection is made you will receive the "L" prompt. At the prompt, press Return, type a period ("."), and press Return again. To complete the connection, type "Dn" (where "n" is the system number) and press Return. You are now ready to logon.

Tymnet. For Tymnet access, dial your local Tymnet number and wait for connection. At the "PLEASE TYPE IN YOUR TERMINAL IDENTIFIER" prompt, type "A," but do not press Return. Then at the "PLEASE LOGIN" prompt, type "DIALCOM;n" (where "n" is the system number) and press Return. You may now logon.

Telenet. For Telenet access, dial your local Telenet number and wait for connection. Press Return twice. When the "TERMINAL=" prompt appears, type "D1" and press Return. Then at the "@" symbol type "C 301 3n" (where "n" is the system number). The numbers you enter here may not be the same as those in this example. Your System Manager will tell you what numbers to use. Your connection is complete and you may now logon.

Datapac. For Datapac access, dial your local Datapac number, and when connected, type two periods ("..") and press Return. You are then prompted for the system address. For example, for the US system address via Telenet, you might type "1 3110 201004444," Datapac's address for system 44. The address will not appear on the screen when typed. You are now ready to logon.

LOGON PROCEDURES

Before logging on to Dialcom, you will need your user ID and password. At the ">" prompt, type "ID," a space, your ID number, and press Return. Next you will be prompted for your password. Type your password and press Return.

>ID ZYX987 (Return)
Password?REDHEAD (Return)

The Electric Mailbox

If your entries are correct, you will see the Dialcom prompt (">"). You are now ready to begin your Dialcom Mail adventure.

USING DIALCOM MAIL

MAIL is a computer-based service that allows you to create, send, read, and file mail messages electronically. As a user, you have a mailbox associated with your ID. Each time you logoff or logon the system, your mailbox is checked automatically to see if you have mail waiting. And if there is mail, you receive a "mail call" and are informed of the number of messages in your mailbox. For example, "Mail call(3 Unread)" indicates that you have three unread messages in your mailbox.

You may check the contents of your mailbox by typing "MAILCK" at any system prompt and pressing Return. You will receive a mail call count of mail in your mailbox.

Commands may be entered in upper- or lower-case, and most commands may be abbreviated. Throughout this chapter, the abbreviation of commands will be shown so you can become familiar with them. Always press Return after entering a command.

Getting Help

Dialcom provides an excellent online tutorial called LEARN. LEARN is a self-instructional, computer-based training program that tells you everything you ever wanted to know about Dialcom's MAIL but were afraid to ask. It is developed in two phases: Basic and Intermediate. "Basic" focuses on sending, reading, scanning and filing electronic mail, while "Intermediate" shows short cuts or command line modes of sending, reading, and scanning mail. Editing capabilities are also discussed. To enter the LEARN tutorial, type "LEARN" at the command prompt.

Help may also be obtained online by typing "HELP" at any command prompt. To receive information on a specific subject, type "HELP" followed by the name of the command with which you need more information. For example, type "HELP MAIL" for information on using electronic mail.

Of course, help is always available by calling the support number, 1-800-435-7342, or by contacting your System Manager or Marketing Support Representative.

Dialcom

Special Keys

Using the following control keys will help make your online experience more enjoyable.

CTRL H	Backspace.
CTRL S	Suspends scrolling of text.
CTRL Q	Resumes scrolling of text.
CTRL P	Terminates the current program. The Break and INTRPT keys perform the same function.
*	Erases an entire line of typed information.

Dialcom has a CHAT feature that permits you to have online conversations with other users. You are notified that someone wants to chat by the other user's ID appearing on your screen followed, perhaps, by a query as to your availability. Chats with others can be interesting and fun, but not if you are busy trying to send mail. There is a way to suppress CHAT. Just type "REFUSE CHAT" at any system prompt and press Return. You will not be interrupted during the current session unless you re-engage CHAT by typing "REFUSE NO CHAT." Check the "Chat" section in the User Guide for complete details on using this feature.

Let's get down to business now and learn how to use Dialcom's electronic mail service, called MAIL. To enter the mail system, type "MAIL" at the system prompt and press Return. You will receive the Mail prompt "Send, Read or Scan:" at which you may choose to send a message, read messages in your mailbox, or check the scan table to see what messages you have waiting. For now, let's learn how to send a letter.

SENDING A LETTER

To send a letter, type "SEND" (S) at the Mail prompt. Then at the "To:" prompt type the ID or directory name of the individual to whom you want to send the message and press Return. Next, you will be asked for the subject of the message. Try to keep the subject short, no more than one line. Type the subject and press Return. The Text prompt is generated. Type your message exactly like you want it to appear to your recipient.

To create a blank line, press Return on an empty line, and be sure to press Return at the end of each line. The message can be any length, but each line should contain no more than 132 characters. Otherwise, the message will be lost. To end the message and send it, type ".SEND" (.S) on a line by itself. Always be sure to type the period before the command, since this tells Dialcom that you are not entering text.

The Electric Mailbox

```
Send, Read or Scan: SEND
To:T.WATSON
Subject:Help
Text:

Dear Watson,
I need you.
Give me a ring asap.

Alex
.Send
```

Notice that you can send a letter using a directory name rather than an ID. Dialcom maintains a mail directory for your organization that contains both the ID number and the name associated with it. To view the directory, type "DIS DIR" at the Mail prompt. The directory shows the name, ID, title, and department for each user.

```
B.BEAGLE  XYZ010  CHIEF HOUND  INVESTIGATIONS
```

The same message may be sent to more than one user, and user names and IDs may be combined on the same line. To do this, type the ID or directory name of each recipient, leaving a space between each entry.

```
To:J.Simit XYZ000 P.Pauper R.Rascal YZX001
```

If you are sending to a long list, end each line with an ampersand (&) to signify to the system that more names follow. Then continue entering names and IDs. Press Return to end the list. You may enter up to 500 names and IDs following the "To:" prompt.

```
To:J.Simit  P.Pauper  YYY002  R.Rascal &
H.Holtz  Y.Young  Z.Ziller &
M.Moose  XZY002
```

Before a letter is sent, it can be edited using the Dialcom Text Editor. Additionally, you can check and correct spelling using the Spell program, or justify the right margin using the Right Justify command. These features are described in detail in the User Guide.

Send Options

Several options are available when sending mail. These options may be entered at the "To:" prompt or as dot commands. As dot commands, they

Dialcom

may be entered at any point in or following the text, but must be entered prior to typing the Send command. As the name implies, a dot command has a dot, or rather a period, before it. So, for example, to use the Courtesty Copy command within the text, you need to type ".CC" on a line by itself followed by the nare or ID of the recipient. The fact that the line begins with a period indicates that a command follows. Be sure none of your text lines begin with a period; otherwise, your text will be taken as commands.

CC   Courtesy Copy. Sends a copy of your message to one or more users. Follow this command by the user ID or directory name of the individual to whom a copy is to be sent. More than one ID or name may be entered, but each must be separated by a space.

BC   Blind Copy. Sends a copy of your message to one or more users. The list of recipients is shown only to the sender and to the one who receives the blind copy. Follow this command by the user ID or directory name of the individual to whom a blind copy is to be sent. More than one ID or name may be entered, but each must be separated by a space.

AR   Acknowledgement Requested. Provides confirmation that the recipient has read your message.

RR   Reply Requested. Generates the Text prompt on the recipient's screen enabling an immediate reply.

EX   Express Priority. Places your message ahead of the recipient's other mail messages in the scan table.

QUIET   Quiet. Prevents display of ID or directory name verifications that normally appears following a Send entry.

NOSHOW   Noshow. Suppresses a display of the list of recipients when a message is read.

Q   Quit. Cancels the letter you are writing. Typing "Quit" at the Mail prompt exits the mail system and takes you to the Dialcom system prompt.

Several dot command options are available for use at the Text prompt or while composing a message. They must, however, be used prior to using the Send (.S) command.

The Electric Mailbox

TO id　　Adds IDs to your list of recipients. Follow the command by a space and one or more ID numbers.

TO -id　　Deletes IDs from your existing list of recipients. Follow the command by a space and one or more IDs.

SU　　Subject. Changes the subject line. Follow this command with a new subject. The subject may not exceed one line.

DIS　　Display. Displays your letter as you have typed it so far.

DIS SU　　Display Subject. Displays the subject line.

DIS HE　　Display Header. Displays both the "To" list and "Subject" title.

DIS DIR　　Display Directory. Displays your mail directory. Press Break to return to the More Text prompt. Type a name with a question mark on each side to search the directory for an ID.

　　　　.DIS DIR ?JONES?

DIS REF　　Display Reference Directories. Displays the directory of your Mail.Ref file and any other files set up for you by your System Manager.

DIS FILES　　Display files. Displays the categories of filed mail messages in your Mail.File.

PA id　　Password. Renders a message unreadable unless the recipient enters a predetermined password.

Messages may also be date activated. There may be times when you want to send a message, but want to delay its delivery until some future date or time. It's a great method to use to send yourself a reminder of an important meeting. The following formats are used.

　　da m/d/y
　　da m/d/y/ h:m

Enter the command at the "To:" prompt following the ID or directory name, or as a dot command in the text of the message.

　　To:j.smith r.ross da 1/4/86/ 11:30

Dialcom

When you are ready to mail your message, type ".SEND" on a line by itself. You will receive a display indicating that the message has been sent. This is followed by a new "To:" prompt. If you don't want to continue composing letters, just type "QUIT" to return to the system prompt, or press Return to go to Mail prompt.

USING MAILING LISTS

Rather than spending time typing names and IDs following the "To:" prompt, it is more cost effective and time efficient to customize your mail directory by preparing mailing lists (called distribution lists by Dialcom) ahead of time.

Mailing lists can be created with the Dialcom Text Editor. To create a list, type "ED" at the system prompt and press Return. You will be prompted for "INPUT." As your first entry on the line, type a name for your list. Then type the IDs and names you want to include on the list, separating each name or ID with a space. The list may contain up to 500 names and IDs. Press Return at the end of each line. To begin a new line of names, first type an ampersand (&), then type more names. Mail options such as CC and EX may also be included in your list. When the list is complete, press Return on a line by itself to generate the Edit prompt. At the prompt, type "SAVE MAIL.REF." That's it. Your list is complete and ready to use.

```
>Edit
INPUT
Marketing b.bolls zzz043 r.smith ex a.baldwin m.jossepher
& cc p.rosenburg d.young zzx900
(Return)
EDIT
Save Mail.Ref
```

To use your mailing list, just type the name you gave the list at the "To:" prompt. Your message will be mailed automatically to everyone on the list.

```
To:Marketing
```

Dialcom also permits you to create alternate directory names for users. For example, rather than sending a message to A.Baldwin, you might prefer to use the name "Anita" each time. To do this, enter "Anita" in your Mail.Ref file as an alternate name. Your entry would look like this:

```
Anita A.Baldwin
```

The Electric Mailbox

Now, whenever you want to send a message to Anita, just type "ANITA" at the "To:" prompt. Dialcom Mail will do the rest.

UPLOADING FILES TO DIALCOM

Dialcom provides several methods for transferring files: PCMAIL, XMIT, and FT. Let's take a look at each of these services.

PCMAIL

PCMAIL is a batch mail transfer service that permits you to create mail offline on you own computer or word processor and transmit it from your disk to another user's mailbox. You may send a single message or several messages at once. This is a great time-saver, and a money-saver, too, especially for prime-time users and foreign users who must pay higher communications surcharges. It is recommended that you send "clean" text, that is text devoid of special formatting codes. If your correspondent's software uses the identical formatting codes as your program, you may send formatted text.

To use PCMAIL, prepare the message on your system just as you would prepare it if you were creating it online. You may use any of the options available online such as CC, BC, and EX. When you finish typing your text, end it with the ".S" (Send) command, press Return, and then (and this is important) type ".END" to indicate to PCMAIL that this is the end of the mail message file. If preparing more than one message, only type ".END" following the last message.

To send the message, logon to Dialcom in the usual way, and at the command prompt type "PCMAIL" and press Return. The PCMAIL banner will be displayed followed by the prompt "Prepare your diskette/cassette and begin sending." At this point, take whatever action your software requires to send the message. When the transfer is complete, you will receive the "Transfer Complete" message, the number of correct messages sent, and the number of incorrect messages (messages that could not be delivered). When all messages have been sent to their recipients, you receive the "All Done" banner and are returned to the system level.

If you made a mistake, such as an incorrect ID, Dialcom will create a file called "PCERROR" in which it will list the errors made. To review this file, type "PCERROR" at the system prompt. Once you know what the mistake is, you can edit it on your word processor and use PCMAIL again to resend it. To learn of other features available when using PCMAIL, review the "File Transfer Services" section of your User Guide.

XMIT

XMIT is a file transfer program that allows you to transfer ASCII files to or from your computer or word processor and your Dialcom Services ID. To transfer a file to your Dialcom ID, logon the system in the usual way and, at the system prompt, type "XMIT." You will be be asked if you want transmit to or from the computer. Type "TO." You are then prompted for the name of the file you want to send. Type the filename and then take whatever action is required by your software to send a file. When the transfer is complete, press Break, or type ".END". If you prefer, you may type ".END" as the last line of your textfile.

To transfer a file from your Dialcom ID to your computer, proceed as above, except type "FROM" at the "Transmit To or From" prompt. Then take whatever action is required by your software to receive a file. When the transfer is complete, press Return. You will be returned to the system level.

FT

Dialcom's FT file transfer service permits you to upload or download files between you PC and your Dialcom service ID. Both ASCII and binary files may be transmitted. This service can be used with any software that supports the XMODEM protocol.

To use this service, logon the system in the usual way and at the system prompt, type "FT." A menu listing various communication software packages is displayed from which you may choose your particular software, if listed. Otherwise, select item 8, "XMODEM." (Dialcom will support any software that has the XMODEM protocol, even if it is not listed on the menu.)

The system then displays the following series of questions:

> Do you want to upload (send to Dialcom) or download (receive from Dialcom) (U,D)?
> Single or batched file protocol (S,B)?
> Are you using CRC (C) or Checksum (K) error detection (C,K)?
> Store data as expanded hex (E), binary (non-mailable) (Z), ASCII (A) or text (T) (E,Z,A,T)?

Enter the appropriate response to each question. You will be prompted for a filename. Type the filename, press Return, and then take whatever action is required by your software to send the file. When transmission is

The Electric Mailbox

complete, press Return to go back to the system level. Refer to your **User Guide** for detailed information on uploading and downloading files.

READING MAIL

To read mail, type "READ" at the Mail "Send, Read or Scan" prompt. The header of the message will be displayed indicating who sent the message, the date and time it was sent, the number of lines, and the subject line. The Read command generates all mail currently in your mailbox. Following the header information, you receive the "--More--" prompt, which tells you that more text follows. To continue reading, press Return. Type "NO" if you don't want to continue. You receive the More prompt after every 23 lines of text until you reach the end of the message. At the end of the message, or if you type "NO" at a More prompt, you receive the "Disposition" prompt.

The Disposition prompt is asking you how you want to handle the just-read message. If you do not want to take immediate action, just press Return. The message will be returned to your mailbox. The following options are available at the Disposition and More prompts.

R Reply. Enables you to reply immediately to the message. The Text prompt is automatically displayed. Type your reply and end it by typing ".SEND." All the Send options discussed earlier are available when you reply to a message.

R All Reply All. Sends a reply to all individuals who are listed after the "To:" prompt in the just-read message.

AP R Append Reply. Permits you to reply to the sender of the message and append the just-read message to the reply.

FO id Forward. Forwards the message to another user. Permits you to add comments to the bottom of the message. The "To:" prompt is displayed followed by the Comments prompt. Enter your comments and end by typing ".SEND." All Send options are available for use with the Forward option.

AP FO id Append Forward ID. Forwards the message to another user and places your comments above the original message.

A Again. Redisplays the text of the just-read message.

Dialcom

A HE Again header. Redisplays the header of the just-read message.

Read Rereads the just-read message beginning with the header.

D Delete. Erases the just-read message from your mailbox.

DQ Delete and Quit. Deletes the message from your mailbox and quits. You are returned to the system level.

Q Quit. You are taken to the Mail prompt. Typing "Quit" at the Mail prompt allows you to exit the Dialcom MAIL service.

In addition to these commands, there are ways you can selectively read messages. Let's review a few of the options.

R UN Read Unread. Reads only unread mail.

R EX Read Express. Reads only express mail.

R UN EX Read Unread Express. Reads only unread express mail.

R F id Read From. Reads only mail from the user identified in the command.

To request mail using one of these options, enter the option following the Mail prompt.

    Send, Read or Scan:R EX

Another way you can read messages is to first check your mailbox using the Scan command. This allows you to view the waiting mail and, if you want, to read mail selectively. The scan table displays the header of each mail message. Each header is preceded by a scan number that can be used to ask to read specific messages. Following the last summary, you receive the Read or Scan prompt. You may view the scan table again or read any or all of the messages.

To read all messages, type "READ" at the prompt. If you want to read only certain mail, type "READ" followed by the scan number of the message you want to view. Numbers may be entered on a single line, but be certain to leave a space between each number. Or a range of numbers may be entered, separated by a hyphen.

The Electric Mailbox

Read or Scan:READ 1 3 5-8 10 15

You may also type, for example, "5-" to read all messages from 5 forward, or "-8" to read all messages up to and including message 8. Still another option is to use the "Rest" command to read the rest of your messages. For example, "READ 3-6 8-19 REST" would read messages in the following order: 3-6, 8-19, 1, 2, and 7.

If you see mail in your mailbox that you don't want to read, simply type "DELETE" followed by the scan number. You may use any of the above options for entering the scan numbers.

Messages that have been read, but not filed or deleted, will be placed, automatically, in your "mail.file" after 30 days. Likewise, unread mail left in your mailbox more than 60 days will be filed in your "mail.file." Your User Guide will show you other ways to selectively read your mail. See the section below for more information on files.

Earlier we talked about sending password protected mail. If you receive a message that is password protected, you will be prompted following the "Subject" to enter a password. If you don't know the password or have forgotten it, you will be unable to read the message. If your correspondent has forgotten the password, the message must be deleted.

FILING, HOLDING, OR SAVING MAIL

Filing Mail

As a Dialcom user, you have a special file associated with your ID. It is referred to as MAIL.FILE. To make location of filed messages easier to find, you may create subfiles within your Mail.File so that mail can be catagorized. To file mail, at the "Disposition" prompt, type "FILE" followed by the name you want to give the file. The category name may not exceed 32 characters, must not begin with a number, and should contain no blank spaces.

Disposition:File Production.Figures

If you have a message that you want to file, but do not want to create a special category for it, just type "FILE" at the Disposition prompt. The message will be filed in a miscellaneous category called "Box." (You don't have to type "box" following the File command.) The following commands may be used to work with your files.

Dialcom

DIS FILES. Display files. Displays categories of files.

SCAN FILE categoryname. Scans filed mail in the named category.

R FILE categoryname. Displays the named file.

D #. Delete Message number. Deletes specified message within a file. This command is used after the Dis Files command.

D FILE categoryname. Deletes an entire category of filed mail.

Holding Mail

There may be occasions when you want to hold a created message for later delivery. This can be done by using the Hold command. To use this command, prepare your message in the usual way, but instead of using the Send command (.S), type ".HOLD" on a line by itself. You receive confirmation that the message has been held instead of sent. The following commands may be used to manipulate your held message.

SC HELD. Scan held. Displays a list of held messages.

S HELD #. Send held message number. Sends the message identified by the message number.

ED HELD #. Edit held message number. Permits you to edit the text of the message using the text editor.

CH HELD #. Change held message number. Permits you to edit the text, To:, and Subject fields of your message. Editing is done using the .SU and .TO commands.

Saving Mail

To save mail that you are sending, at the Text prompt type ".SAVE" followed by a filename you want to give the message. To save mail that you have read, at the Disposition prompt type "SAVE" followed by the file name you want to give the message.

SHORTCUTS

As you become familiar with the commands and prompts available in Dialcom Mail, you will probably want to use options when typing MAIL in the command mode. This allows you to enter all your commands on one line without waiting for individual prompts.

Command mode is easy to use. Simply type the commands on one line, separating each command with a space. Then press Return to end the line. For example, to send a message to J.Pepper with a subject title of "Sales Contest Results," and to send a blind copy to T.Salt, you would make the following entry:

>Mail j.pepper bc t.salt su sales contest results

If the command line is long, end each line with an ampersand (&) to tell the system that you have not finished entering names and IDs. Remember, too, that the subject (SU) command must be the last command in the line.

OTHER SERVICES

XMAIL

Dialcom XMAIL is used to send telexes and telegrams. This service literally puts the world at your fingertips. To enter XMAIL, type "XMAIL" at the system prompt. You will receive the Send or Queue prompt, the primary prompt in XMAIL. For all services discussed below, refer to the User Guide for complete details.

Sending A Telex I Message

Enter the XMAIL service as described above and, at the Send or Queue prompt, type "SEND." Then at the "To:" prompt, type "TLX" followed by the telex address to which the message is to be sent. Telex I addresses consist of up to eight digits.

To: TLX34-000111

Next, you are prompted for the text of your message. Each line of text may consist of up to 68 characters. If you exceed this number, you will receive an error message and will have to edit the too-long line. When the message is complete type ".SEND" and press Return. You will be informed of the number of words it contains and then receive the "Message Queued" message. Your telex is queued for being sent electronically to ITT's switching center in New York called "Timetran." When the message is actually delivered, you receive confirmation in your mailbox.

Dialcom

Sending A Telex II Message

Telex II messages are prepared much the same way as Telex I messages. At the "To:" prompt, type "TWX followed by the 10-digit Telex address. All Telex II addresses begin with 501, 610, 710, 810, or 910, which denotes the area of the United States or Canada to which the message is being sent.

>To: TWX 501-000-0001

The telex number may be entered with or without hyphens. Type the subject of your message at the Subject prompt and then enter the text. When the message is complete, type ".SEND" on a line by itself and press Return.

Sending An International Telex

An international telex is sent to a Telex I printer and is prepared similar to other Telex messages. But in addition to the address, you must also include a three-digit "Country Code" immediately preceding the telex number. This tells XMAIL the name of the country to which the telex is being sent. If you are unsure of the proper code, at the "To" prompt type "DIS" followed by the name of the country to which you are sending the telex. You will be advised of the appropriate code.

>To: DIS Great Britain

>Great Britain 851

At the next "To" prompt, type "ITX," a space, the country code, and then the telex address of your recipient.

>To: ITX 851276900221(ABTCOMM)

Notice that an "answerback" has been included in the address. An answerback is used to enable Timetran to verify that your message was sent to the right address. (Answerbacks may be used with all telex messages as an optional entry.)

Another optional feature is the "Attention line." An attention line is similar to the "subject" line on a normal message. Up to 10 attention lines may be entered. The attention line is always enclosed in single quotes. If the attention line consists of more than one name, separate each name with a slash (/). And if more than one attention line is needed, end each line with an ampersand (&), press Return, and continue the addresses on the following line. You will be prompted for "More" addresses.

   To: ITX 85127690021(ABTCOMM)'Sandy Lake/Suzy Sichel &
   More:Rocky Rhodes/Dusty Fields'

Certain characters are not permitted when preparing international telexes. Other characters such as the dollar sign will appear as different characters. Check your User Guide for a complete list of these.

Sending A Speedmail Letter

Speedmail letters are transmitted to teleprinters located at ITT Speedmail Delivery Centers where they are printed, placed in envelopes, and then sent through the US Postal Service. Speedmail may be sent anywhere in the US, except Hawaii, and Canada. To send a Speedmail Letter, type "SPD" at the "To:" prompt, followed by the name, address, city, state and zip of your correspondent. The address may consist of up to five lines, with up to 33 characters per line. The address is formatted as follows:

     To:SPD 'name/address/city,state zip'

When typing the "To" information, enter the slash (/) marks and single quotes just as they are shown in the above format. In addition, if the address requires the entry of more than one line, type an ampersand (&) at the end of each line, except the last line, where you must press Return.

     To:SPD 'Mr. Joseph P. Wall/Manager/ABT Communications/ &
     More:1234 West Main St./Omaha, NE 50123'

If any part of the address is missing, you will receive an error message and you will need to re-enter the address before the letter can be sent.

When prompted for text, enter your message just as you want it to appear to your correspondent. Each line may contain no more than 68 characters, including spaces. The options available at the Text prompt are the same as those available in Mail.

When sending Speedmail, you must also provide a return mailing address. This is entered following the text of your message, before issuing the Send command. It is formatted exactly like the recipient's address, except that you must type ".MYADDRESS" before entering the actual address.

Dialcom

    .myaddress 'name/address/city,state zip'

Check your User Guide to learn how your return address can be filed in your PARAM.INI file for future use. When all appropriate entries have been made, type ".SEND" to send your Speedmail on its way. When your letter has been received by Timetran, you will receive a confirmation in your mailbox.

Sending A Mailgram

Western Union Mailgrams are prepared in a similar way to Speedmail Letters. The main differences are that you type "MLG" before the address entry, and a return address is optional. If you do not specify a return address, the address of the Mailgram center that processed the message will be used.

    To:MLG 'name/address/city, state zip'

Remember that each address line may consist of up to 33 characters and you may include up to 5 address lines. Each line in the text may contain up to 68 characters, including spaces. Your Mailgram is routed to Western Union and then sent via the US or Canadian post office to the recipient.

Sending a Cablegram or Lettergram

The methods used to send cablegrams and telegrams are similar. Line lengths are limited to 68 characters, and the message is ended by typing ".SEND" on a line by itself.

For cablegrams, at the "To:" prompt type "ICB" followed by the name, address, city, country, and zip code. Enclose the name of the country in parentheses, and type single quotes on each side of the address. Be certain to separate each line of the address with a slash (/).

    To:ICB 'name/address/city,(country)zip'

Your cablegram is sent to a teleprinter in the specified city and country's local telephone company and then delivered to the recipient, usually within several hours.

For lettergrams, at the "To:" prompt type "ILT." Then complete the address the same as for cablegrams.

    To:ILT 'name/address/city,(country)zip'

The Electric Mailbox

Your lettergram is sent to a teleprinter in the specified city and country's local post office and then mailed to the recipient the following business day.

Sending Options

There are many sending options available when using Dialcom's XMAIL. For a complete review of these, study the "General XMAIL Options" section in your User Guide.

LOGOFF PROCEDURES

To logoff the system, simply type "OFF" at the system prompt. You will receive a display indicating the time and date of your logoff, the amount of computer time in seconds you used, and the number of connect minutes. The "@" symbol following the disconnect message means that you are still connected to the network and could, if you wanted, logon again. To disconnect your terminal from the network, you must turn off your modem or hang up the telephone.

CONCLUSION

Dialcom Mail is a user-friendly system that is easy to learn and to use. And it works with any equipment you already have. The LEARN tutorial is an asset to any new or occasional user. Dialcom offers options such as the Spell Program and text justification, which are not available on other systems. Because of the minimum charges, the system is best suited for medium to large organizations. However, if you want a service that provides "one stop shopping" where all your electronic mail needs can be met efficiently, Dialcom is well worth your consideration.

Appendices

Network Access Numbers

Tymnet

*2400 baud +300 baud

Alabama

Anniston	205-236-2655
Birmingham	205-942-4141
	205-942-7898*
Dothan	205-792-0914
Florence	205-767-7131
Huntsville	205-882-3003
Mobile	205-343-8414
Montgomery	205-265-4570
Tuscalossa	205-349-5670

Alaska

Ancorage	907-338-7222
Cold Bay	907-532-2371
Cordova	907-424-3744
Dead Horse	907-659-2777
Delta Junction	907-895-5070
Fairbanks	907-456-3282
Glennallen	907-822-5231
Homer	907-235-5239
Juneau	907-789-7009
Kenai	907-262-1990
King Salmon	907-246-3049
Kotzebue	907-442-2602
Palmer/Wasilla	907-745-0200
Prudhoe Bay	907-659-2777
Soldotna	907-262-1990
Wadilla	907-745-0200

Arizona

Mesa	602-254-5811
	602-258-0554*
Phoenix	602-254-5811
	602-258-0554*
Tuscon	602-790-0764
	602-747-4856*

Arkansas

Fayetteville	501-442-3850
Fort Smith	501-782-2486
Hot Springs	501-321-9741
Jonesboro	501-935-7957
Little Rock	501-666-6886
Pine Bluff	501-535-8055
Springdale	501-442-3850

California

Alameda	415-430-2900
	415-633-1896*
Alhambra	818-308-1800
Anaheim	714-756-8341
	714-852-8141*
Antioch	415-754-8222
Arcadia	818-308-1800
Bakersfield	805-322-5078
Belmont	415-366-1092
	415-361-8701*
Berkeley	415-430-2900
	415-633-1896*
Beverly Hills	818-789-9002
Burbank	818-846-9235
Burlingame	415-952-4757
Canoga Park	818-789-9002
Cathedral City	916-324-0920
Chico	916-893-1876
Colton	714-370-1200
Concord	415-685-6003
Corona	714-371-2291
Covina	714-594-4567
Davis	916-758-3551
Diamond Bar	714-594-4567
El Monte	818-308-1800
El Segundo	213-640-1281
	213-643-4388*
Escondido	619-941-6700
Eureka	707-444-0491
Fairfield	707-447-3436
Fremont	415-490-7366
Fresno	209-442-4328
Glendale	818-846-9235
Hayward	415-430-2900
	415-633-1896*
Inglewood	213-587-0030
	213-587-7514*
Irvine	714-756-8341
	714-852-8141*
Lancaster	805-948-4602
Long Beach	213-435-0900
Los Altos	408-980-8100
	408-986-0646*
Los Angeles	213-587-0030
	213-587-7514*
Manteca	805-985-7843
Mar Vista	213-643-2907
	213-643-4388*
Marina del Rey	213-643-2907
	213-643-4388*
Merced	209-384-7227
Modesto	209-527-0150
Monterey	408-372-6433
Napa	707-257-2656
Newport Beach	714-756-8341
	714-852-8141*
Norwalk	213-435-0900

Oakland	415-430-2900	Boulder	303-830-9210
	415-633-1896*		303-832-3447*
Ontario	714-594-4567	Colorado Springs	303-590-1003
Oxnard	805-985-7843	Copper Mountain	303-968-6560
Pacheco	415-685-6003	Denver	303-830-9210
Palm Springs	619-324-0920		303-832-3447*
Palo Alto	415-366-1092	Fort Collins	303-221-0678
	415-361-8701*	Greeley	303-325-0960
Pasadena	818-308-1800	Pueblo	303-543-3313
Pleasant Hill	415-685-6003		
Pleasanton	415-462-8900	Connecticut	
Pomona	714-594-4567		
Rancho Bernardo	619-485-1990	Bloomfield	203-242-7140
Redding	916-241-4820		203-242-1984*
Redwood City	415-366-1092	Bridgeport	203-367-6021
	415-361-8701*	Danbury	203-797-9539
Riverside	714-370-1200	Fairfield	203-226-5250
Sacramento	916-448-4300	Hartford	203-242-7140
	916-447-7434*		203-242-1986*
Salinas	408-757-0147	Meriden	203-634-9249
San Bernadino	714-370-1200	Middletown	203-634-9249
San Clemente	714-492-0165	New Haven	203-773-0082
San Diego	619-296-3370	New London	203-443-6997
	619-296-8747*	North Haven	203-773-0082
San Fernando	818-789-9002	Norwalk	203-444-1709
San Francisco	415-974-1300	Norwich	203-444-1709
	415-543-0691*	Stamford	203-965-0000
San Jose	408-980-8100		203-327-2974*
	408-986-0646*	Stratford	203-367-6021
San Luis Obispo	805-546-8541	Waterbury	203-755-5994
San Mateo	415-952-4757	Westport	203-226-5250
San Pedro	213-435-0900		
San Rafael	415-492-9320	Deleware	
Santa Ana	714-756-8341		
	714-852-8141*	Dover	302-678-0449
Santa Barbara	805-963-8731	Newark	302-652-2060
	805-963-9241	Wilmington	302-652-2060
Santa Clara	408-980-8100		
	408-986-0646*	District of Columbia	703-691-8200
Santa Cruz	408-475-0981		703-691-9390
Santa Monica	213-643-2907		
	213-643-4388*	Florida	
Santa Rosa	707-527-6180		
Sherman Oaks	818-789-9002	Boca Raton	305-272-7900
Stockton	209-467-0601	Boyton Beach	305-471-9310
Sunnyvale	408-980-8100	Clearwater	813-796-2166
	408-986-0646*	Daytona Beach	904-255-4783
Thousand Oaks	805-496-3473	Delray Beach	305-272-7900
Vallejo	707-644-1192	Fort Lauderdale	305-463-0882
Van Nuys	818-789-9002		305-467-1870*
Ventura	805-985-7843	Fort Myers	813-481-8866
Vernon	213-587-0030	Fort Pierce	305-466-0661
	213-587-7514*	Gainesville	904-378-4576
Visalia	209-625-5523	Hollywood	305-624-7900
Vista	619-941-6700		305-624-0304*
Walnut Creek	415-938-9550	Jacksonville	904-721-8100
West Covina	714-594-4567		904-721-8559*
West Los Angeles	818-709-9002	Lakeland	813-665-8582
Woodland	916-758-3551	Longwood	305-841-0020
			305-841-0217*
Colorado		Melbourne	305-676-4336
		Merritt Island	305-459-0671
Aurora	303-830-9210		

Miami	305-624-7900	Libertyville	312-362-0820
	305-624-0304*	Maywood	312-345-9100
Ocala	904-351-0305	Palatine	312-358-8770+
Orlando	305-841-0020	Peoria	309-637-5961+
	305-841-0217*	Rock Island	309-788-3713
Panama City	904-769-9446	Rockford	815-398-6090
Pensacola	904-477-3344	Springfield	217-525-8025
Pompano Beach	305-272-7900	Saint Charles	312-844-1800
Sarasota	813-365-6980	Urbana	217-359-1163
St. Petersburg	813-796-2166	Wheaton	312-790-4400
Tallahassee	904-878-2267		
Tampa	813-932-7070	Indiana	
	813-933-6210*		
West Palm Beach	305-471-9310	Bloomington	812-332-3051
Winterhaven	813-665-8582	Evansville	812-464-8181
		Fort Wayne	219-422-2581+
Georgia		Gary	219-885-0002
		Hammond	219-885-0002
Albany	912-883-2246	Highland	219-885-0002
Athens	404-548-7006	Indianapolis	317-631-1002
Atlanta	404-446-0270		317-632-6408*
	404-446-1508*	Kokomo	317-456-3871
Augusta	404-868-8500	Layfayette	317-423-4544
Columbus	404-327-0597	Marion	317-664-9033
Macon	912-744-0605	Mishawaka	219-234-5005
Marietta	404-446-0270	Muncie	317-284-4474
	404-446-1508*	South Bend	219-234-5005+
Marinez	404-868-8500	Terre Haute	812-232-3605
Norcross	404-446-1508		
	404-446-0270	Iowa	
Rome	404-291-1000		
Savannah	912-232-6751	Cedar Falls	319-236-9020
Warner Robins	912-746-2739	Cedar Rapids	319-363-7514
		Davenport	319-788-3713
Hawaii		Des Moines	515-277-7752
		Dubuque	319-582-3599
Honolulu	808-545-7610	Iowa City	319-354-7371
		Marshalltown	515-753-0670
Idaho		Sioux City	712-255-3834
		Waterloo	319-236-9020
Boise	208-343-0404		
Idaho Falls	208-523-7800	Kansas	
Pocatello	208-233-2501		
		Kansas City	913-384-1544
Illinois		Lawrence	913-843-4870
		Leavenworth	913-682-2660
Aurora	312-844-1800	Manhattan	913-776-0121
Belleville	618-277-9806	Mission	913-384-1544
Bloomington	309-829-2802		913-384-0055*
Champaign	217-359-1163	Salina	913-823-7186
Chicago	312-922-4601	Shawnee Mission	913-384-1544
	312-922-6571*	Topeka	913-234-3070
Cicero	312-345-9100	Wichita	316-681-0832
Danville	217-432-3133		316-681-2719*
Decatur	217-429-5995		
Downers Grove	312-790-4400	Kentucky	
Forest Park	312-345-9100		
Freeport	815-233-5585	Bowling Green	502-781-5711
Glen Ellyn	312-790-4400	Lexington	606-253-3463
Joliet	815-727-7669	Louisville	502-499-7110
Kankakee	815-932-0850	Owensboro	502-685-1318
Lake Forest	312-362-0820		
Lake Zurich	312-358-8770+		

Louisiana

Alexandria	318-445-2694
Baton Rouge	504-924-5102
Lafayette	318-237-9500
Lake Charles	318-494-1991
Monroe	318-322-4109
New Orleans	504-522-1370
	504-525-2014*
Shreveport	318-688-5840
Slidell	504-649-6050

Maine

Auburn	207-786-5390
Bangor	207-989-2831
Brewer	207-989-2831
Lewiston	207-786-5390
Portland	207-775-5971

Maryland

Aberdeen	301-273-7100
Annapolis	301-263-2344
Baltimore	301-547-8100
	301-528-9296*
Chevy Chase	301-652-0800
Cumberland	301-777-9320
Frederick	301-293-3380
Hagerstown	301-293-3380
Myersville	301-293-3380
Rockville	301-652-0800

Massachusetts

Attleboro	617-226-4471
Boston	617-292-1900
	617-357-5052*
Brockton	617-584-6873
Cambridge	617-292-1900
	6170357-5052*
Fall River	617-675-1750
Fitchburg	617-537-6451
Framingham	617-620-1264
Holyoke	413-781-6830
Lawrence	617-689-0186
Leominser	617-537-6451
Lowell	617-459-0694
Lynn	617-537-6451
New Bedford	617-599-0015
Pittsfield	413-499-0971
Salem	617-599-0015
Springfield	413-781-6830
Taunton	617-824-6692
Worburn	617-935-2057
Worcester	617-791-9000

Michigan

Ann Arbor	313-662-8282
Battle Creek	616-962-1851
Benton Harbor	616-925-3134
Cadillac	616-775-6089
Detroit	313-962-2870
	313-963-3460*
Flint	313-238-8917
Freeland	517-695-6751
Grand Rapids	616-459-2304
Jackson	517-782-0584
Kalamazoo	616-388-2130
Lansing	517-482-5721
Manistee	616-723-6296
Midland	517-695-6751
Muskegon	616-725-8136
Plymouth	313-451-2400
Port Huron	313-982-0301
Saginaw	517-695-6751
Saint Joseph	616-925-3134
Southfield	313-424-8024
Traverse City	616-947-0050

Minnesota

Duluth	218-722-7441
Mankato	507-625-9481
Minneapolis	612-333-2799
	612-332-4024*
Rochester	507-289-5599
St. Cloud	612-252-9093
St. Paul	612-333-2799
	612-322-4024*

Mississippi

Gulfport	601-868-0109
Meridian	601-693-8216
Pascagoula	601-769-6673
	601-769-6502+
Vicksburg	601-638-1551

Missouri

Bridgeton	314-731-8002
	314-731-8283*
Columbia	314-874-2771
Jefferson City	314-634-8296
Joplin	417-782-3037
Kansas City	913-384-1544
	913-384-0055*
Rolla	314-364-2034
Saint Joseph	816-232-1897
Saint Louis	314-731-8002
	314-731-8283*
Springfield	417-869-4903

Montana

Billings	406-252-4880
Bozeman	406-587-1930
Butte	406-782-4202
Great Falls	406-727-9510
Missoula	406-721-5965

Nebraska

Lincoln	402-474-7481
Omaha	402-393-0903
	402-393-0982*

Nevada

Boulder City	702-293-0300
Carson City	702-885-8411
Las Vegas	702-293-0300
Reno	702-885-8411

New Hampshire

Durham	603-868-1502
Hanover	603-643-1457
Manchester	603-623-0409
Nashua	603-882-0435
Salem	603-893-6200

New Jersey

Atlantic City	609-345-4050
Camden	609-665-5600
	609-665-5902*
Cherry Hill	609-665-5600
	609-665-5902*
Eatontown	201-758-9100
Elizabeth	201-758-9100
	201-824-3044*
Englewood Cliffs	201-567-9841
	201-567-8951*
Fair Lawn	201-567-9841
	201-567-8951*
Jersey City	201-824-1212
	201-824-3044*
Long Branch	201-758-9100
Lyndhurst	201-460-0180
	201-460-0100+
Morristown	201-539-1222
Newark	201-824-1212
Park Ridge	201-742-0752
Patterson	201-742-0752
Pennsauken	609-665-5600
Piscataway	201-562-9700
Princeton	609-452-1018
Rahway	201-499-4747
Red Bank	201-758-9100
Ridgewood	201-742-0752
Trenton	609-888-0803
Union	201-824-1212
	201-824-3044*
Wayne	201-742-0752

New Mexico

Albuquerque	505-242-8344
Las Cruces	505-525-0011
Santa Fe	505-984-2941

New York

Albany	518-458-8300
Binghamton	607-724-4351
Buffalo	716-845-6610+
	716-852-1077*
Corning	607-962-4481
Elmira	607-737-9065
Hempstead	516-485-7422
Huntington	516-420-1221
Ithaca	607-257-6601
Melville	516-420-1221
Mineola	516-485-7422
New York	209-625-5523
	914-638-0882
	212-809-9660*
Niagara Falls	716-285-2561
Perington	716-385-5817
	716-385-5710*
Pittsford	716-385-5817
	716-385-5817*
Poughkeepsie	914-473-0401
Rochester	716-385-5817
	716-385-5710*
Ronkonkoma	516-467-5178
Syracuse	315-437-7111
Utica	315-797-7001
White Plains	914-328-7730
	914-761-9590*

North Carolina

Asheville	704-252-9023
Chapel Hill	919-549-8952
Charlotte	704-252-9023
	704-377-0521
	704-374-0803*
Durham	919-549-8952
Fayetteville	919-485-8161
Greensboro	919-273-0332
Greenville	919-758-7854
High Point	919-884-4364
Raleigh	919-829-0536
Rocky Mount	919-549-8952
Wilmington	919-762-1865
Winston-Salem	919-784-6080

North Dakota

Bismarck	701-255-0869
Fargo	701-280-0210
Grand Forks	701-746-0344
Minot	701-838-1114

Ohio

Akron	216-535-1861
Canton	216-455-0066
Cincinnati	513-530-9019
	513-530-9021*
Cleveland	216-241-0024
	216-861-6709*
Columbus	614-221-1862
	614-221-1612*
Dayton	513-223-3847
Hamilton	513-894-1521

Lima	419-228-6343	Rhode Island	
Mansfield	419-526-6067	Middletown	401-849-1660
Marysville	513-644-3896	Newport	401-849-1660
Newark	614-349-9234	Pawtucket	401-273-0200
North Canton	216-455-0066	Providence	401-273-0200
Springfield	513-324-3816		401-247-7380*
Steubenville	614-283-2496	Warwick	401-273-0200
Toledo	419-255-7790	Woonsocket	401-765-2400
Warren	216-394-6529		
Youngstown	216-759-8892	South Carolina	
		Charleston	803-556-1040
Oklahoma		Columbia	803-254-7563
		Greenville	803-271-9213
Ardmore	405-223-1552	Spartanburg	803-585-0016
Enid	405-234-5320		
Lawton	405-353-6987	South Dakota	
Oklahoma City	405-495-8201	Rapid City	605-341-4007
	405-495-9201*	Sioux Falls	605-335-0780
Tulsa	918-582-4433		
	918-582-9653*	Tennessee	
		Chattanooga	615-265-1020
Oregon		Clarkesville	615-552-0193
		Johnson City	615-928-1191
Eugene	503-746-0092	Knoxville	615-690-1543
Medford	503-779-2902		615-889-5790
Portland	503-222-0900	Memphis	901-527-8006
	503-222-2151*	Nashville	615-885-3530
Salem	503-585-0182	Oak Ridge	615-482-9080
Springfield	503-746-0092		
		Texas	
Pennsylvania			
		Abilene	915-672-1363
Allentown	215-865-6978	Amarillo	806-371-0031
Altoona	814-943-5848	Arlington	817-877-3630
Bethlehem	215-865-6978	Austin	512-444-3280
Butler	412-285-9387	Baytown	713-422-9746
	412-285-9527	Brownsville	512-548-1331
Coatsville	215-383-0440	Bryan	409-823-1090
Downingtown	215-383-0440	College Station	409-823-1090
Erie	814-456-8501+	Corpus Christi	512-883-8050
Greensburg	412-836-4470	Dallas	214-638-8888
Harrisburg	717-763-6481		214-630-5516*
King of Prussia	215-337-9900	Denton	817-565-0552
Lancaster	717-397-7731	El Paso	915-533-1453
Latrobe	412-836-4470	Fort Worth	214-877-3630
Levittown	215-943-3700	Galveston	409-765-7338
New Castle	412-654-1530	Houston	713-556-6700
Norristown	215-666-9190		713-496-1332*
Philadelphia	215-567-4390	Killeen	817-634-3439
	215-557-9903*	Laredo	512-727-4626
Pittsburgh	412-642-6778	Longview	214-236-4271
	412-642-2015	Lubbock	806-797-0765
Reading	215-779-9580	McAllen	512-631-6101
Scranton	717-346-4516	Midland	915-683-5645
State College	814-234-3853	Nederland	409-722-4166
Valley Forge	215-666-9190	Odessa	915-563-3745+
Wilkes Barre	717-826-8991	Port Arthur	409-722-4166
York	717-846-3900	San Angelo	915-658-6697
		San Antonio	512-225-8002
Puerto Rico			512-222-9877*
		Sherman	214-893-1024
San Juan	809-792-5900	Texas City	409-765-7338

Tyler	214-597-9321
Waco	817-753-0001
Wichita Falls	817-723-2386

Utah

Ogden	801-621-1280
Provo	801-375-0645
Salt Lake City	801-364-0780
	801-533-8152*

Vermont

Barre	802-229-4508
Burlington	802-864-5714
Montpelier	802-229-4508

Virginia

Alexandria	703-691-8200
	703-325-3136*
Arlington	703-691-8200
Charlottesville	804-293-5714
Fairfax	703-691-8200
	703-352-3136*
Hampton	804-596-7608
Lynchburg	804-528-1903
Midlothian	804-744-4860
Newport News	804-596-7608
Norfolk	804-855-7751
Petersburg	804-861-1788
Portsmouth	804-855-7751
Richmond	804-744-4860
	804-744-5039
Roanoke	703-344-2762
Virginia Beach	804-855-7751
Williamsburg	804-872-9592

Washington

Auburn	206-825-7720
Bellevue	206-285-0109
	206-281-7141*
Enumclaw	206-825-7720
Everett	206-258-1018
Olympia	206-943-9050
Richland	509-375-3367
Seattle	206-285-0109
Spokane	509-624-1549
Tacoma	206-272-1503
Vancouver	206-693-0371
Yakima	509-248-1462

West Virginia

Charleston	304-345-9575
Huntington	304-523-8432+
Morgantown	304-292-0682
Parkersburg	304-428-8511
Westover	304-292-0682

Wisconsin

Appleton	414-722-5580
Beloit	608-365-6883
Brookfield	414-785-1614
	414-785-0630*
Eau Claire	715-832-1345
Green Bay	414-432-3964
Jamesville	608-365-6883
La Crosse	608-784-9099
Madison	608-221-0891
Milwaukee	414-785-1614
	414-785-0630*
Neenah	414-722-5580
Oshkosh	414-235-7473
Racine	414-632-3006
Sheboygan	414-452-2622
West Bend	414-334-1240
Wausau	715-355-1262

Wyoming

Casper	307-234-4211
Cheyenne	307-632-7464

CANADA

British Columbia

Burnaby	604-294-3223
Vancouver	604-294-3223

Ontario

Ottawa	613-563-7703

Quebec

Montreal	514-748-1051
Ville St. Lauren	514-748-1051

Telenet

*2400 baud #1200 baud +300 baud

Alabama

Bessemer	205-328-2310
Birmingham	205-328-2310
	205-251-1885*
Florence	205-767-7960
Huntsville	205-539-2281
Mobile	205-432-1680
	205-438-6881*
Montgomery	205-269-0090
Sheffield	205-767-7960

Arizona

Mesa	602-254-0244
Phoenix	602-254-0244
	602-256-6955*
Scottsdale	602-254-0244
Tempe	602-254-0244
Tucson	602-747-0107

Arkansas

| Little Rock | 501-372-4616 |

California

Alhambra	818-507-0909
Anaheim	714-558-7078
Bakersfield	805-327-8146
Burlingame	415-591-0726
Canoga Park	213-306-2984
Colton	714-824-9000
Compton	213-516-1007
Concord	415-827-3960
Covina	818-330-1630
Cupertino	408-294-9119
Danville	415-829-6705
Escondido	619-741-7756
El Monte	818-507-0909
Fullerton	714-558-7078
Fresno	209-233-0961
Garden Grove	714-898-9820
Glendale	818-507-0909
	818-246-3886*
Hayward	415-881-1382
Hollywood	213-624-2251
	213-937-3580
Huntington Bch	714-558-7078
Inglewood	213-624-2251
	213-937-3580
Los Angeles	213-624-2251
	213-937-3580
	213-622-1138*
Los Altos	415-856-9995
Long Beach	213-548-6141
Marina del Rey	213-306-2984
	213-823-0173*
Modesto	209-576-2852
Monterey	408-375-2675
Mountain View	415-856-9995
Newport Beach	714-558-7078
Norwalk	213-404-2237
Oakland	415-836-4911
	415-834-3194*
Oceanside	619-430-0613
Oxnard	805-656-6760
Palo Alto	415-856-9995
Pasadena	213-507-0909
Redwood City	415-591-0726
Riverside	714-824-9000
Sacramento	916-448-6262
	916-443-7434*
Salinas	408-443-4940
San Bernadino	714-824-9000
San Carlos	415-591-0726
	415-595-8870*
San Diego	619-233-0233
	619-231-1703*
San Francisco	415-362-6200+
	415-956-5777
	415-788-0825*
San Jose	408-294-9119
	408-286-6340*
San Mateo	415-591-0726
San Pedro	213-548-6141
San Rafael	415-492-0752
San Ramone	415-829-6705
Santa Ana	714-558-7078
	714-550-4625*
Santa Barbara	805-682-5361
Santa Clara	408-294-9119
Santa Cruz	408-429-6937
	408-429-6937*
Santa Monica	213-306-2984
Santa Rosa	707-578-4447
Stockton	209-473-2056
Sunnyvale	408-294-9119
Torrance	213-548-6141
Walnut Creek	415-856-9995
Woodland Hills	818-887-3160
	818-348-7141*
Woodside	415-856-9995
Ventura	805-656-6760

Colorado

Aurora	303-337-6060
	303-696-0159*
Boulder	303-337-6060
Colorado Springs	303-635-5361
Denver	303-337-6060
Fort Collins	303-493-9131
Lakewood	303-337-6060
Pueblo	303-542-4053

Connecticut

Bridgeport	203-335-5055
	203-367-9130*
Danbury	203-794-9075
Greenwich	203-348-0787
Hartford	203-247-9479
	203-724-9396*
Milford	203-624-5954
New Haven	203-624-5954
	203-773-3569*
Stamford	203-348-0787
	203-359-9404*
Waterbury	203-753-4512
West Hartford	203-247-9479

Delaware

Wilmington	302-454-7710
	302-737-4340*

District of Columbia

Washington	202-429-7896+
	202-429-7800#
	202-429-0956*

Florida

Boca Raton	305-368-8300
Clearwater	813-323-4026
Daytona Beach	904-252-9914
Ft. Lauderdale	305-764-4505
	305-524-5304*
Ft. Myers	813-337-0308
Gainesville	904-377-3005
Holly Hill	904-255-2629
Jacksonville	904-353-1818
	904-791-9201*
Lakeland	813-688-4366
Melbourne	305-676-1393
Miami	305-372-0230
	305-372-1355*
Ocala	904-351-3790
Orlando	305-422-4088
	305-422-8858*
Pensacola	904-432-1335
Pompano Beach	305-941-5445
St. Petersburg	813-323-4026
	813-327-1163*
Sarasota	813-923-4563
Tallahassee	904-681-1902
Tampa	813-224-9920
	813-223-5960*
W. Palm Beach	305-833-6691

Georgia

Albany	912-883-8600
Athens	404-549-4524
Atlanta	404-523-0834
	404-584-0212*
Augusta	404-724-2752
Columbus	404-571-0556
Macon	912-741-1011
Savannah	912-236-2605

Hawaii

Honolulu	808-524-8110+
	808-524-8380#

Idaho

Boise	208-343-0611
Lewistown	208-743-0099

Illinois

Arlington Hts	312-938-0500+
	312-938-0600#
Aurora	312-859-8483
Bloomington	309-827-7000#
Champaign	217-384-6428
	217-328-0317*
Chicago	312-938-0500+
	312-938-0600#
	312-938-8725*
Cicero	312-938-0500+
	312-938-0600#
Decatur	217-422-0835
DeKalb	815-758-2623
East St. Louis	314-421-4990
Joliet	815-726-0070
Oak Park	312-938-0500+
	312-938-0600#
Peoria	309-637-8570
Rockford	815-965-0400
Skokie	312-938-0500+
	312-938-0600#
Springfield	217-753-1373
Urbana	217-384-6428

Indiana

Bloomington	812-332-1344
Evansville	812-424-7693
Ft. Wayne	219-426-2268
Gary	219-882-8800
Indianapolis	317-634-5708
	317-299-6766*
Kokomo	317-455-2460
Lafayette	317-742-1165
Mishawka	219-233-7104
Munice	317-289-5068

Osceola	219-233-7104	Cambridge	617-292-0662
South Bend	219-233-7104	Chicopee	413-781-3811
Terre Haute	812-234-8429	Farmington	617-879-6798
		Holyoke	413-781-3811
Iowa		Lexington	617-863-1550
		Lowell	617-937-5214
Cedar Rapids	319-364-0911	Medford	617-292-0662
Council Bluffs	402-341-7733	New Bedford	617-999-2915
Davenport	319-324-2445	Newton	617-292-0662
Des Moines	515-288-4403	Quincy	617-292-0662
Iowa City	319-351-1421	Somerville	617-292-0662
Sioux City	712-255-1545	Springfield	413-781-3811
		Waltham	617-292-0662
Kansas		Wood Hole	617-540-7500
		Worcester	617-755-4740
Kansas City	816-221-9900		
Topeka	913-233-9880	Michigan	
Wichita	316-262-5669		
		Ann Arbor	313-996-5995
Kentucky			313-665-2900*
		Battle Creek	616-968-0929
Bowling Green	502-782-7941	Columbia	314-449-7947
Frankfort	502-875-4654	Detroit	313-964-2988
Lexington	606-233-0312		313-963-2274*
Louisville	502-589-5580	Flint	313-235-8517
	502-583-1006*	Grand Rapids	616-774-0966
		Jacksom	517-782-8111
Louisiana		Kalamazoo	616-345-3088+
		Lansing	517-484-0062
Baton Rouge	504-343-0753	Muskegon	616-726-5723
Lafayette	318-233-0002	Saginaw	517-790-5166
Lake Charles	318-436-0518	Southfield	313-827-4710
Monroe	318-387-6330	Traverse City	616-946-2121
New Orleans	504-524-4094	Warren	313-575-9152
	504-522-3967*		
Shreveport	318-221-5833	Minnesota	
		Duluth	281-722-1719
Maine		Minneapolis	621-341-2459
			612-338-1661*
Augusta	207-622-3123	Rochester	507-282-5917
Lewistown	207-784-0105	St. Paul	621-341-2459
Portland	207-773-4219		
		Mississippi	
Maryland			
		Jackson	601-969-0036
Annapolis	301-224-8550		
Baltimore	301-727-6060	Missouri	
	301-752-2837*		
Bethesda	202-429-7896+	Florissant	314-421-4990
	202-429-7800#	Jefferson City	314-634-5178
Dundalk	301-727-6060	Kansas City	816-221-9900
Rockville	202-429-7896+		816-472-1430*
	202-429-7800#	St. Joseph	816-279-4797
Silver Spring	202-429-7896+	St. Louis	314-421-4990
	202-429-7800#		314-421-0381*
Towson	301-727-6060	Springfield	417-886-0531
Massachusetts		Montana	
Arlington	617-292-0662		
Boston	617-292-0662	Billings	406-245-7649
	617-574-9244*	Helena	406-443-0000
Brookline	617-292-0662	Misscula	406-721-5900

Nebraska

Lincoln	402-475-4964
Omaha	402-341-7733

Nevada

Las Vegas	702-737-6861
Reno	702-827-6900

New Hampshire

Concord	603-224-1024
Manchester	603-668-1420
Nashua	603-889-8618
Portsmouth	603-431-2302

New Jersey

Atlantic City	609-348-0561
Bayonne	201-623-0469
Jersey City	201-623-6818+
	201-623-0469
Marlton	609-596-1500
Morristown	201-455-0275
	201-644-4745*
New Brunswick	201-745-2900
	201-745-7010*
Newark	201-623-0469
	201-623-7122*
Passaic	201-778-5600
	201-773-3674*
Patterson	201-684-7560
	201-742-4415*
Princeton	609-799-5587
Trenton	609-989-8847
Union City	201-623-0469

New Mexico

Albuquerque	505-243-4479
Santa Fe	505-473-3403

New York

Albany	518-465-8444
	518-465-8632*
Binghamton	607-772-6642
Buffalo	716-847-1440
Deer Park	516-667-5566
Hempstead	516-292-3800
	516-485-3380*
Ithaca	607-257-3227
New York	212-620-6000
	212-741-4950
	212-741-8100
	212-645-0560*
Plattsburgh	518-562-1890
Poughkeepsie	914-473-2240
Rochester	716-454-1020
	716-454-5730*
Schenectady	518-465-8444

Syracuse	315-472-5583
	315-479-5445*
Troy	518-465-8444
Utica/Rome	315-797-0920
White Plains	914-328-9199
	914-682-3505*

North Carolina

Asheville	704-252-9134
Charlotte	704-332-3131
	704-333-6204*
Davidson	919-549-8139
Durham	919-549-8139
Fayetteville	919-323-8165
Greensboro	919-273-2851
High Point	919-899-2253
Raleigh	919-549-8139
Research Tri.	919-549-8139
	919-541-9096*
Wilmington	919-343-8773
Winston-Salem	919-725-2126

North Dakota

Fargo	701-237-3442
Mandan	701-663-2256

Ohio

Akron	216-678-5115
Canton	216-452-0903
Cincinnati	513-579-0390
	513-241-8008*
Cleveland	216-575-1658
	216-771-6480*
Columbus	614-463-9340
	614-451-5573
	614-461-9044*
Dayton	513-461-5254
	513-461-0755*
Euclid	216-575-1658
Elyria	216-323-5059
Kent	216-678-5115
Mansfield	419-526-0686
Parma	216-575-1658
Springfield	513-324-1520
Toledo	419-255-7881
Youngstown	216-743-1296
Warren	216-394-0041

Oklahoma

Bethany	405-232-4546
Norman	405-232-4546
Oklahoma City	405-232-4546
Stillwater	405-624-1112
Tulsa	918-584-3247

Oregon

Corvallis	503-754-9273
Eugene	503-683-1460

Medford	503-779-6343		512-929-3622*
Portland	503-295-3028	Bryan	409-779-0173
	503-241-0496*	Corpus Christi	512-884-9030
Salem	503-378-7712	Dallas	214-748-0127+
			214-748-6371
Pennsylvania			214-745-1359*
		El Paso	915-532-7907
Allentown	215-435-3330	Fort Worth	817-332-4307
Erie	814-899-2241		817-332-6794*
Harrisburg	717-236-6882	Galveston	409-762-4382
	717-236-2007*	Houston	713-227-1018
Johnstown	814-535-7576		713-227-8208
King of Prussia	215-337-4300	Lackland	512-225-8004
	215-337-2850*	Laredo	512-724-1791
Lancaster	717-393-2154	Longview	214-236-3196
Penn Hills	412-288-9950+	Lubbock	806-747-4121
	412-288-9974#	Midland	915-561-9811
Philadelphia	215-574-9462		915-561-8597*
	215-574-0990*	Nederland	409-722-3720
Pittsburgh	412-288-9950+	Odessa	915-561-9811
	412-288-9974#	San Angelo	915-944-7621
	412-471-6430*	San Antonio	512-225-8004
Reading	215-372-7116		512-225-3444*
Scranton	717-961-5321	Temple	817-773-9423
State College	814-231-1510	Terminal	915-561-9811
Upper Darby	215-574-9462	Tyler	214-592-3927
Williamsport	717-494-1796	Waco	817-752-9743
York	717-846-6550		
		Utah	
Rhode Island			
		Ogden	801-627-1630
Providence	401-751-7912	Provo	801-373-0542
Warwick	401-751-7912	Salt Lake City	801-359-0149
			801-359-0578*
South Carolina			
		Vermont	
Charleston	803-722-4303		
Columbia	803-254-0695	Burlington	802-864-0808
Greenville	803-233-2486	Montpelier	802-229-4966
Spartanburg	803-585-1637		
		Virginia	
South Dakota			
		Alexandria	202-429-7896+
Pierre	605-244-0481		202-429-7800#
		Allandale	202-429-7896+
Tennessee			202-429-7800
		Chesapeake	804-625-1186
Bristol	615-968-1130	Fairfax	202-429-7896+
Chattanooga	615-756-1161		202-429-7800#
	615-265-7929*	Falls Church	202-429-7896+
Knoxville	615-523-5500		202-429-7800#
	615-521-5072*	Harrisonburg	703-434-7121
Memphis	901-521-0215	Herndon	703-435-1800
	901-527-5175*	Newport News	804-596-6600
Nashville	615-244-3702	Norfolk	804-625-1186
	615-255-2608		804-625-2408*
		Portsmouth	804-625-1186
Texas		Richmond	804-788-9902
			804-353-0219*
Abilene	915-676-9151	Roanoke	703-334-2036
Amarillo	806-373-0458	Springfield	202-429-7896+
	806-373-1833*		202-429-7800#
Austin	512-928-1130		

Vienna	202-429-7896+
	202-429-7800#
Virginia Beach	804-625-1186

Washington

Auburn	206-939-9982
Bellevue	206-625-9612
Bellingham	206-733-2720
Longview	206-577-5835+
Olympia	206-754-0460
Richland	509-375-5377
Seattle	206-625-9612
	206-623-9951*
Spokane	509-455-4071
Tacoma	206-627-1791
Wenatchee	509-663-6227

West Virginia

Charleston	304-345-6471
Huntington	304-523-2802
Morgantown	304-292-0104

Wisconsin

Eau Claire	715-836-9295
Green Bay	414-432-2815
Madison	608-257-5010
Milwaukee	414-271-3914
	414-278-8007*
Neenah	414-722-7636
Racine	414-552-7217
Sheboygan	414-452-3995

Wyoming

Casper	307-265-5167
Cheyenne	307-638-4421

TERMINAL IDENTIFIERS

Terminal	Type
Adds Consul 520, 580, 980	D1
Adds Envoy 620, Regent	D1
Alanthus Data Terminal T-133	A1
Alanthus Data Terminal T-300	A8
Alanthus Data Terminal T-1200	A3
Alanthus Miniterm	A2
AM-Jacquard Amtex 425	D1
Anderson Jacobsen 510	D1
Anderson Jocobsen 630	B1
Anderson Jocobsen 830, 832	B3
Anderson Jocobsen 841 codes	IBM 2741
Anderson Jocobsen 860	B5
Apple II	D1
Atari 400, 800	D1
AT&T Dataspeed 40/1, 40/2, 40/3	D1
Beehive Minibee, Microbee	D1
Centronics 761	A8
Commodore Pet	D1
Compu-Color II	D1
Computer Devices CDI 1030	A2
Computer Devices Teleterm 1132	A8
Computer Devices Teleterm 1200 Series	A2
Computer Transceiver Execuport 300	A2
Computer Transceiver Execuport 1200	A9
Computer Transceiver Execuport 4000	A8
CPT 6000, 8000	D1
Datamedia Elite	D1
Datapoint 1500, 1800, 2200, 3000, 3600, 3800	D1
Data Products Portaterm	A1
Data Terminal and Communications DTC 300, 302	B3
Diablo Hyterm	B3
Digi-Log 33 & Telecomputer II	D1
Digital Equipment (LA 35-36) Decwriter II	A8
(LA 120) Decwriter III	A8
Digital Equipment VT50, VT52, VT100, WS78, WS200	D1
Gen-Comm Systems 300	B3
GE Terminet 30	A5
GE Terminet 300	A4
GE terminet 120, 1200	A3
General Terminal GT-100-A, GT-101, GT-110, GT-400, GT-400B	D1
Hazeltine 1400, 1500, 2000	D1
Hewlett Packard 2621	D3
Hewlett Packard 2640 Series	D1
IBM 2741 (EBCD Code) Typesphere Element Code:	
#963, 996, 998	E1
938, 939, 961, 997	E2
942, 943	E3
947, 948	E4
IBM 2741 (Correspondence Code) Typesphere Element Code:	
#001, 005, 007, 008, 022, 030, 050, 053, 067, 070, 085	C1
006, 010, 015, 019, 059, 090	C2
021, 025-029, 031-039, 060 068, 086, 123, 129-145, 156, 161	C3
043, 054	C4
IBM 3101	D1
Informer 1304, D304	D1
Infoton 100, 200, 400 Vistar	D1
Intelligent Systems Intecolor	D1
Interec Intertube II	D1
Lanier Word Processor	D1
Lear Siegler ADM Series	D1
Lexitron 1202, 1303	D1
Memorex 1240	A2
Micom 2000, 2001	D1

NBI 3000	D1
NCR 260	A2
Perkin-Elmer Model 1100, Owl, Bantam	D1
Perkin-Elmer Carousel 300 Series	A8
Radio Shack TRS 80	D1
Research Inc. Teleray	D1
Tektronix 40002-4024	D1
Teletype Model 33, 35	A1
Teletype Model 40	D1
Teletype Model 43	B3
Teletype Model 40/1, 40/2, 40/3	D1
Texas Instruments 725	A7
Texas Instruments 733	A2
Texas Instruments 735	A6
Texas Instruments 743, 745, 763, 765	D1
Texas Instruments 820	B3
Texas Instruments 99/4	D1
Trendata 1000, 1500, 2000 codes	IBM 2741
Trendata 4000 (ASCII)	B1
Tymshare 110, 212	A2
Tymshare 315	A8
Tymshare 325	B3
Univac DCT 500	B4
Wang 20, 25, 30, 015, 130, 145	D1
Western Union EDT 33, 35	A1
Western Union EDT 300	A3
Western Union 1200	A4
Xerox 800, 850, 860	A4
Xerox 1700	B3

Uninet

Alabama
  Bessemer   205-324-5440
  Birmingham   205-324-5440
  Mobile   205-432-5590
  Montgomery   205-265-9210

Arizona
  Glendale   602-253-1940
  Mesa   602-253-1940
  Phoenix   602-253-1940
  Scottsdale   602-253-1940
  Tempe   602-253-1940
  Tuscon   602-747-7868

Arkansas
  Little Rock   501-372-5098

California
  Alhambra   213-748-0203
  Anaheim   714-553-1740
  Arcadia   213-748-0203
  Belmont   415-592-4140
  Berkeley   415-398-7533
  Beverly Hills   213-748-0203
  Burbank   213-748-0203
  Canoga Park   213-748-0203
  Colton   714-359-5732
  Compton   213-435-9295
  Concord   415-674-0721
  Covina   714-623-6088
  Downey   213-435-9295
  El Monte   213-748-0203
  El Segundo   213-215-3690
  Fremont   415-965-2701
  Fresno   209-225-1354
  Fullerton   714-553-1740
  Glendale   213-748-0203
  Hayward   415-581-4582
  Hollywood   213-748-0203
  Huntington Bch   714-553-1740
  Inglewood   213-215-3690
  Long Beach   213-435-9295
  Los Altos   415-965-2701
  Los Angeles   213-748-0203
  Mar Vista   213-748-0203
  Monrovia   213-748-0203
  Monterey   408-373-1601
  Mountain View   415-965-2701
  Newport Beach   714-553-1740
  Norwalk   213-435-9295
  Oakland   415-652-3566
  Ontario   714-623-6088
  Oxnard   805-487-4971
  Palo Alto   415-965-2701
  Pasadena   213-748-0203
  Pomona   714-623-6088
  Redwood City   415-592-4140
  Richmond   415-652-3566
  Riverside   714-359-5732
  Sacramento   916-443-2472
  San Carlos   415-592-4140
  San Diego   619-458-1250
  San Francisco   415-398-7533
  San Jose   408-293-9767
  San Leandro   415-652-3566
  San Mateo   415-592-4140
  San Pedro   213-435-9295
  Santa Ana   714-553-1740
  Santa Clara   408-293-9767
  Santa Monica   213-215-3690
  Stockton   209-462-9981
  Sunnyvale   408-293-9767
  Torrance   213-215-3690
  Vallejo   707-554-0910
  Van Nuys   213-748-0203
  Ventura   805-487-4971
  Walnut Creek   415-674-0721

Colorado
  Arvada   303-740-8649
  Aurora   303-740-8649
  Boulder   303-740-8649
  Colorado Spgs   303-636-5104
  Denver   303-740-8649
  Grand Junction   303-241-2911
  Lakewood   303-740-8649

Connecticut
  Bloomfield   203-247-7723
  Bridgeport   203-367-7476
  Danbury   203-794-9298
  Greenwich   203-325-4414
  Hartford   203-247-7723
  Milford   203-367-7476
  New Britain   203-247-7723
  New Haven   203-777-8376
  New London   203-777-8376
  Stamford   203-325-4414
  Waterbury   203-574-1924
  West Hartford   203-247-7723

Deleware
  Wilmington   302-999-8915

District of Columbia
  Washington   202-347-3337

Florida
  Boca Raton   305-368-4245
  Cape Canaveral   305-768-0031

Clearwater	813-821-7561		Wichita	316-264-9505
Daytona Beach	904-257-1071		**Kentucky**	
Fort Lauderdale	305-467-6504		Lexington	606-254-4475
Fort Myers	813-337-3323		Louisville	502-589-5837
Gainesville	904-373-0522		**Louisiana**	
Hialeah	305-591-1254		Baton Rouge	504-387-5294
Hollywood	305-467-6504		Lafayette	318-232-4743
Jacksonville	904-356-1115		Metairie	504-834-7113
Melbourne	305-768-0031		New Orleans	504-834-7113
Miami	305-591-1254		Shreveport	318-222-7810
Miami Beach	305-591-1254		**Maine**	
Naples	813-263-4522		Augusta	207-623-4065
Orlando	305-894-4815		Bangor	207-947-5261
Sarasota	813-365-3028		**Maryland**	
St. Petersburg	813-576-0280		Baltimore	301-366-3102
Tallahassee	904-681-0486		Bethesda	202-347-3337
Tampa	813-273-9160		Dundalk	301-366-3102
West Palm Beach	305-659-6107		Rockville	202-347-3337
Georgia			Silver Spring	202-347-3337
Atlanta	404-252-0999		Towson	301-366-3102
Marietta	404-252-0999		**Massachusetts**	
Savannah	912-233-9117		Andover	617-475-4012
Illinois			Arlington	617-890-1808
Arlington Hts	312-663-9600		Boston	617-890-1808
Aurora	312-663-9600		Brookline	617-890-1808
Bloomington	309-827-4681		Cambridge	617-890-1808
Champaign	217-356-0827		Holyoke	413-734-6447
Chicago	816-474-1129		Lexington	617-890-1808
East St. Louis	314-878-7705		Lynn	617-890-1808
Elgin	312-663-9600		Medford	617-890-1808
Evanston	312-663-9600		Newton	617-890-1808
Joliet	815-727-0951		Northampton	413-586-4941
Oak Park	321-663-9600		Quincy	617-890-1808
Peoria	309-673-9103		Somerville	617-890-1808
Rock Island	309-788-0871		Springfield	413-734-6447
Rockford	815-965-4240		Waltham	617-890-1808
Skokie	312-663-9600		Worcester	617-791-9752
Springfield	217-544-0728		**Michigan**	
Urbana	217-356-0827		Ann Arbor	313-761-5402
Waukegan	312-663-9600		Dearborn	313-581-4844
Indiana			Detroit	313-884-7017
Evansville	812-424-0045		Flint	313-767-1102
Fort Wayne	219-422-2491		Grand Rapids	616-774-0270
Indianapolis	317-236-9638		Grosse Pointe	313-884-7017
Mishawaka	219-674-6963		Jackson	517-782-9335
Osceola	219-674-6963		Kalamazoo	616-388-7300
South Bend	219-674-6963		Lansing	517-484-3116
Iowa			Livonia	313-844-7017
Ames	515-233-3610		Muskegon	616-739-4316
Cedar Rapids	217-356-0827		Plymouth	313-761-5402
Council Bluffs	402-345-4913		Royal Oak	313-581-4844
Davenport	309-788-0871		Saginaw	517-752-1134
Des Moines	515-244-5838		St Clair Shores	313-884-7017
Kansas			Southfield	313-884-7017
Kansas City	816-474-1129		Sterling Hts	313-884-7017
Overland Park	816-474-1129		Taylor	313-884-7017
Shawnee Mission	816-474-1129			
Topeka	913-273-1081			

Warren	313-884-7017	**New York**	
Westland	313-884-7017	Albany	518-449-1772
		Babylon	516-422-5980
Minnesota		Buffalo	716-884-5980
Bloomington	612-377-7041	Cheektowaga	716-884-5980
Minneapolis	612-377-7041	Hicksville	516-933-6200
Saint Cloud	612-253-4751	Huntington	516-351-1431
Saint Paul	612-377-7041	Ithaca	607-272-0211
		Latham	518-499-1772
Missouri		Mineola	516-294-3950
Bridgeton	314-878-7705	Mount Vernon	212-736-3660
Columbia	314-874-4065	New Rochelle	914-328-7844
Creve Coeur	314-878-7705	New York	212-736-3660
Florissant	314-878-7705	Plainview	516-933-6200
Independence	816-474-1129	Rochester	716-454-2510
Jefferson City	314-635-8400	Schenectady	518-449-1772
Kansas City	816-474-1129	Syracuse	315-471-6114
Liberty	816-474-1129	Tonawand	716-884-5980
Saint Louis	314-878-7705	Troy	518-449-1772
		White Plains	914-328-7844
Nebraska		Yonkers	914-328-7844
Lincoln	402-474-7734		
Omaha	402-345-4913	**North Carolina**	
		Chapel Hill	919-929-0054
Nevada		Charlotte	704-365-6630
Las Vegas	702-383-5931	Davidson	919-682-9671
Reno	702-786-7503	Durham	919-682-9671
		Greensboro	919-272-6033
New Hampshire		Raleigh	919-782-3930
Concord	603-224-1336	Research Trl.	919-682-9671
Hanover	603-643-6631	Tarboro	919-823-7503
Merrimack	603-424-6171	Winston-Salem	919-748-0965
New Jersey		**North Dakota**	
Atlantic City	609-345-1001	Bismarck	701-222-4844
Bayonne	201-226-0220		
Branchburg	201-722-2261	**Ohio**	
Camden	609-338-1177	Akron	216-434-0032
Carteret	201-750-9190	Brook Park	216-267-9200
Cherry Hill	609-338-1177	Canton	216-453-2329
East Orange	201-226-0220	Cincinnati	513-621-3433
Edison	201-572-5555	Cleveland	216-267-9200
Elizabeth	201-750-9190	Columbus	614-464-9941
Fords	201-750-9190	Dayton	513-461-0220
Jersey City	201-623-7863	Euclid	216-267-9200
Morristown	201-993-8018	Hamilton	513-621-3433
New Brunswick	201-572-5555	Kent	216-434-0032
Newark	201-623-7863	Parma	216-267-9200
Oradell	201-599-2200	Toledo	419-241-3514
Passaic	201-599-2200		
Paterson	201-599-2200	**Oklahoma**	
Piscataway	201-572-5555	Bartlesville	918-336-9447
Princeton	609-683-4550	Bethany	405-948-8513
River EAdge	201-599-2200	Britton	405-948-8513
Roseland	201-226-0220	Norman	405-948-8513
Somerville	201-722-2261	Oklahoma City	405-948-8513
Trenton	609-393-1930	Tulsa	918-747-2431
Union City	201-599-2200		
Woodbridge	201-750-9190	**Oregon**	
		Hood River	503-386-1129
New Mexico		Portland	503-287-4282
Albuquerque	505-843-7981		

Pennsylvania
  Allentown   215-437-4654
  Bethlehem   215-437-4654
  Blue Bell   215-643-7600
  Carlisle   717-243-4919
  Danville   717-275-1001
  Harrisburg   717-657-4954
  Hershey   717-567-4954
  King of Prussia 215-567-3340
  Lancaster   717-299-6603
  Penn Hills   412-931-9360
  Philadelphia   215-567-3340
  Pittsburgh   412-931-9360
  Reading   215-375-6186
  Upper Darby   215-567-3340
  Valley Forge   215-567-3340
  York   717-843-0965

Rhode Island
  Pawtucket   401-781-8505
  Providence   401-273-7900
  Warwick   401-273-7900

South Carolina
  Columbia   803-256-5248

South Dakota
  Rapid City   605-348-5099
  Sioux Falls   605-338-6061

Tennessee
  Bristol   615-968-5360
  Chattanooga   615-267-2248
  Knoxville   615-523-5611
  Memphis   901-767-9431
  Nashville   615-256-6469

Texas
  Amarillo   806-379-9211
  Arlington   214-438-1256
  Athens   214-677-2798
  Austin   512-473-2617
  Beaumont   409-839-4731
  Bryan   409-822-0143
  College Station 409-822-0143
  Corpus Christi   512-241-5691
  Dallas   214-438-1256
  Fort Worth   214-263-1480
  Garland   214-438-1256
  Grand Prairie   214-263-1480
  Houston   713-552-9659
  Irving   214-438-1256
  Lackland   512-647-0031
  Pasadena   713-552-9659
  Port Authur   409-839-4731
  San Antonio   512-647-0031
  Tyler   214-597-8113

Utah
  Salt Lake City   801-364-8985

Vermont
  Rutland   802-775-0722

Virginia
  Alexandria   202-347-3337
  Annandale   202-347-3337
  Arlington   202-347-3337
  Charlottesville 804-979-4433
  Chesapeake   804-627-1295
  Covington   703-962-2126
  Fairfax   202-347-3337
  Falls Church   202-347-3337
  Hampton   804-244-5700
  Lynchburg   804-846-0943
  Newport News   804-244-5700
  Norfolk   804-627-1295
  Portsmouth   804-627-1295
  Richmond   804-353-3813
  Roanoke   703-345-3246
  Springfield   202-347-3337
  Vienna   202-347-3337
  Virginia Beach   804-627-1295

Washington
  Bellevue   206-641-4210
  Bellingham   206-733-8800
  Everett   206-339-2655
  Seattle   206-641-4210
  Spokane   509-838-7715
  Tacoma   206-627-8778

West Virginia
  Charleston   304-344-1650
  Huntington   304-529-0304

Wisconsin
  Appleton   414-739-1486
  Brookfield   414-549-3970
  Green Bay   414-435-0133
  Madison   608-258-8408
  Milwaukee   414-549-3970
  Racine   414-632-7395
  Sheboygan   414-459-7455
  Sun Prairie   608-258-8408

Wyoming
  Laramie   307-721-4145

CompuServe

*2400 baud #1200 baud +300 baud

Alabama
  Birmingham   205-879-2250+
                                    205-879-2280#
  Bessemer   205-897-2250+
                                    205-879-2280#
  Huntsville   205-536-4405
  Mobile   205-478-0688
  Montgomery   205-262-0010

Arizona
  Mesa   602-256-2951
  Phoenix   602-256-2951
                                    602-267-0623
                                    602-225-0200*
  Scottsdale   602-256-2951
  Tempe   602-256-2951
  Tuscon   602-748-2004+
                                    602-748-2009
  Yuma   602-783-7191

Arkansas
  Little Rock   501-224-9311

California
  Alameda   415-531-3700
  Anaheim   714-520-9733
                                    714-520-9724+
  Bakersfield   805-323-7691
  Berkeley   415-531-3700
  Beverly Hills   213-739-0371
                                    213-739-8906
                                    213-487-6461#
                                    213-383-9284*
  Canoga Park   818-902-0932
                                    818-902-0934
  Castro Valley   415-581-2631
  Cathedral City   619-325-4584
  Concord   415-682-2633
  Culver City   213-216-0001
                                    213-390-9617
                                    213-216-0010
                                    213-397-8812#
  Cupertino   408-988-8762
  Fresno   209-252-1892
  Hayward   415-581-2631

Hollywood   818-982-1813
Inglewood   213-739-0371
                                  213-739-8906
                                  213-487-6461#
                                  213-383-9284*
Irvine   714-851-0145
Livermore   415-443-9202
Long Beach   213-591-8392
Los Altos   408-988-8762
Los Angeles   213-739-0371
                                  213-739-8906
                                  213-487-6461#
                                  213-383-9284*
Monterey   408-375-9930
Mountain View   408-988-8762
Newport Beach   714-851-0145
North Hollywood   818-982-1813
Oakland   415-531-3700
Pacheco   415-682-2633
Palm Springs   619-325-4584
Palo Alto   415-591-5846
                                  415-591-5591+
Pleasant Hills   415-682-2633
Pomona   714-623-2651
Rancho Bernardo   619-471-0960
Riverside   714-359-7801
Sacramento   916-971-4681
San Bernardino   714-881-1583
San Carlos   415-591-5591+
                                  415-591-5846
San Diego   619-283-6091
                                  619-569-0697
                                  619-283-6021+
                                  619-569-8324*
San Francisco   415-956-4281
                                  415-956-4191+
                                  415-956-4281#
                                  415-982-9055#
                                  415-982-9055*
San Fernando   213-739-0371
                                  213-739-8906
                                  213-383-9284
                                  213-487-6461#
San Jose   408-988-8762
San Mateo   415-591-5846
                                  415-591-5591+
                                  415-591-5415*
Santa Clara   408-988-8762
                                  408-988-5366*
Santa Barbara   805-682-2331
Sherman Oaks   818-902-0932+
                                  818-902-0934
Sierra Madre   818-303-2563
                                  818-303-2681
Solana Beach   619-481-3527
Stockton   209-465-7251
Sunnyvale   408-988-8762
                                  408-988-5366*
Thousand Oaks   805-499-0371
                                  805-499-0566
Torrance   213-542-4311
Van Nuys   818-902-0932+
                                  818-902-0934

Ventura	805-643-0177	Ft. Myers	813-939-7060
Walnut Creek	415-682-2633	Jacksonville	904-396-7105
West Los Angeles	213-739-0371	Longwood	305-273-8780
	213-739-8906		305-273-8805
	213-487-6461#	Miami	305-266-0231
	213-383-9284*	Orlando	305-273-8805
Colorado			305-273-8780
Aspen	303-925-5892	Pensacola	904-434-3911
	303-623-4711#	St. Petersburg	813-525-0378
Aurora	303-629-5563+	Sarasota	813-355-9331
	303-337-0668#	Tallahassee	904-222-4144
Boulder	303-629-5563+		904-224-6021
	303-629-0068#	Tampa	813-875-0633
	303-623-4711#	Vero Beach	305-778-0550
Colorado Springs	303-596-0910	West Palm Beach	305-684-9051
Denver	303-629-0068		
	303-629-5563+	Georgia	
	303-623-4711#	Albany	912-435-9420
	303-629-9145*	Atlanta	404-237-3003
Dillon	303-668-0991		404-237-8113+
Ft. Collins	303-493-8601		404-231-3214*
Glenwood Springs	303-945-0424	Augusta	404-733-0346
Grand Junction	303-241-1889	Martinez	404-733-0346
	303-241-1885+		
Lakewood	303-629-5563	Hawaii	
	303-623-4711#	Kailua	808-263-6670
Vail	303-476-8700		
		Idaho	
Connecticut		Boise	208-384-5666
Bridgeport	203-926-0001		208-384-5660+
Danbury	203-797-1815	Pocatello	208-232-9452
Fairfield	203-926-0001		
Greenwich	203-967-4589	Illinois	
Hartford	203-728-0633	Arlington Hts	312-332-7382
Milford	203-926-0001		312-443-1250+
New Haven	203-467-3489		312-372-1402#
New London	203-444-2509		312-263-5636*
Norwalk	203-967-4589	Aurora	312-896-2137
Stamford	203-967-4589	Chicago	312-332-7382
Waterbury	203-574-0500		312-433-1250+
	203-573-0392#		312-372-1402#
Westport	203-222-1748+		312-263-5636*
	203-226-2704	Cicero	312-332-7382
	203-222-1742#		312-443-1250+
			312-372-1402#
Delaware			312-263-5636*
Newark	302-652-8732	East St. Louis	314-241-3101
	302-656-6852#		314-241-3102
	302-656-6451*		314-241-3110#
Wilmington	302-652-8732	Lombard	312-953-9680
	302-656-6852#	Oak Park	312-332-7382
	302-656-6451*		312-433-1250+
			312-372-1402#
District of Columbia	703-352-7500		312-263-5636*
	703-841-9834	Rockford	815-968-3412
	703-352-8750#	St. Charles	312-896-2137
		Skokie	312-332-7382
			312-443-1250
Florida			312-372-1402#
Boyton Beach	305-684-9051		312-263-5636*
Daytona Beach	904-257-5019	Springfield	217-522-5101
Ft. Lauderdale	305-771-8074		
	305-772-3240		

Indiana
  Elkhart   219-293-1593
  Evansville   812-479-0165
  Fort Wayne   219-447-0510
  Gary   219-769-0081
  Indianapolis   317-638-2762
                 317-638-2517+
                 317-638-5785#
  Layfayette   317-742-6578
  Muncie   317-284-3812
  Osceola   219-674-6951
                 219-679-4705#

Iowa
  Cedar Rapids   319-365-9363
  Davenport   319-323-7388
  Des Moines   515-270-9410
                 515-270-1581

Kansas
  Mission   816-474-3770
                 816-472-1283*
  Shawnee   816-474-3770
                 816-472-1283*
  Wichita   316-689-8765
                 316-689-8585

Kentucky
  Lexington   606-259-3446
  Louisville   502-581-9526
                 502-581-9804#

Louisiana
  Baton Rouge   504-273-0184
  Lafayette   318-233-1150
  Monroe   318-387-0879
                 318-325-6781*
  New Orleans   504-734-8150
  Shreveport   318-424-5380

Maine
  Portland   207-879-0005

Maryland
  Annapolis   301-266-7530
  Baltimore   301-254-7113+
                 301-254-1150#
                 301-254-1652*
  Bethesda   703-252-7500
                 703-352-8750#
  Dundalk   301-254-7113
                 301-254-7311+
                 301-254-1150#
                 301-254-1652*
  Hyattsville   301-599-0200
  Towson   301-254-1150#
                 301-254-1652*

Massachusetts
  Arlington   617-542-1796
                 617-542-1779#
  Amherst   413-256-8591
  Boston   617-542-3792+
                 617-542-1796
                 617-542-1779#
                 617-542-7148*
  Brockton   617-588-3222
  Brookline   617-542-1796
                 617-542-1779#
  Burlington   617-229-2740
  Cambridge   617-542-1796
                 617-542-1779#
  Concord   617-371-0354
  Framingham   617-875-3814
  Georgetown   617-352-7596
  Hudson   617-568-8019
  Maynard   617-897-4746
  Medfield   617-359-7603
  Medford   617-542-3792+
                 617-542-1796
                 617-542-1779#
                 617-542-7148*
  Medway   617-533-2772+
  Mendon   617-478-0653+
  Newton   617-542-3792+
                 617-542-1796
                 617-542-1779#
  Springfield   413-734-7362
  Quincy   617-542-3792+
                 617-542-1796
                 617-542-1779#
  Waltham   617-542-3792+
                 617-542-1796
                 617-542-1779#
                 617-542-7148*
  Westboro   617-366-2617
  Worcester   617-792-2512+

Michigan
  Ann Arbor   313-663-3934
  Detroit   313-255-9207
                 313-255-9304*
  Flint   313-238-6202
  Grand Rapids   616-459-9891
  Kalamazoo   616-344-5312
                 616-344-2298+
  Lansing   517-321-2388
  Saginaw   517-893-1161
  Troy   313-362-2540

Minnesota
  Minneapolis   612-342-2207
                 612-375-0328#
                 612-339-2507*
  St. Paul   612-342-2207
                 612-375-0328#
                 612-339-2507*

Mississippi
  Jackson   601-948-6411

Missouri
  Florissant   314-241-3101
                 314-241-3102
                 314-241-3110#

Independence	816-474-3770		212-758-0330#
	816-472-1283*		212-758-2090
Kansas City	816-474-3770		212-969-7790*
	816-472-1283*	Poughkeepsie	914-473-2617
Saint Louis	314-241-3102	Rochester	716-458-3465
	314-241-3101		716-458-3460
	314-241-3110#	Schenectady	518-439-7491
		Syracuse	315-458-6016
Montana		Tonawanda	716-694-6263
Billings	406-245-0863	Troy	518-439-7491
		White Plains	914-428-9270
Nebraska			914-949-4510
Lincoln	402-474-1006	Williston Park	516-294-1482
Omaha	402-895-5288		
	402-896-3853#	North Carolina	
		Burlington	919-584-2971
Nevada		Chapel Hill	919-682-6239
Las Vegas	702-878-0056	Charlotte	704-333-7155
Reno	702-786-5356		704-333-6654+
	702-786-5308+	Davidson	919-725-1550
	702-786-7416#	Durham	919-682-6239
		Greensboro	919-373-1635
New Hampshire		Raleigh	919-878-8570
Nashua	603-883-5551	Research Tri.	919-682-6239
		Winston-Salem	919-725-2126
New Jersey			
Atlantic City	609-645-1258	Ohio	
Bayonne	201-624-6565	Akron	216-867-1243
Camden	609-665-7555		216-867-1237
Cherry Hill	609-665-7555	Athens	614-594-8364
Elizabeth	201-624-6565	Canton	216-455-2126
Greenbrook	201-968-9000		216-455-2516
	201-968-0263*	Cincinnati	513-771-1630
Hackensack	201-489-0111		513-241-2187#
Hackettstown	201-852-8070		513-771-1760#
	201-852-8502	Cleveland	216-771-6860
Jersey City	201-624-6565		216-771-0723+
Montclair	201-783-5400		216-771-8350#
Newark	201-624-6565		513-771-1760#
Parsippany	201-898-1935		216-771-4014*
	201-898-0259#	Columbus	614-457-2105+
Pennsaukin	609-665-7555		614-876-2116
Princeton	609-683-4776		614-451-5573#
	609-683-4770		614-761-1131*
	609-921-8930#	Dayton	513-461-1064
Tom's River	201-244-7722	Euclid	216-771-0723+
Union City	201-624-6565		216-771-6860
Wayne	201-742-0752		216-771-8350#
			216-771-4014*
New Mexico		Granville	614-587-0932
Albuquerque	505-265-1263	North Canton	216-867-1237
	505-265-7046#		216-867-1243
Los Alamos	505-662-4122	Parma	216-771-0723+
			216-771-6860
New York			216-771-8350#
Albany	518-439-7491		216-771-4014*
Buffalo	716-874-3751	Toledo	419-244-0073
Hicksville	516-681-7347		419-244-6286#
	516-681-7240		419-243-2818*
Lake Grove	516-981-0880	Youngstown	216-743-4992
New York	212-758-4114		
	212-422-8820		
	212-344-5674#		

275

Oklahoma
  Bethany   405-946-4799+
                          405-946-4860
  Norman   405-946-4799+
                          405-946-4860
  Oklahoma City   405-946-4860
                          405-946-4799+
  Tulsa   918-749-8850
                          918-749-8801+

Oregon
  Portland   503-232-4026
                          503-232-1072

Pennsylvania
  Allentown   215-776-6960
  Erie   814-453-7538
  Harrisburg   717-657-9633
  King of Prussia   215-279-5811
  Penn Hills   412-391-7732
                          412-391-8818
                          412-391-8218#
                          412-261-4192*
  Philadelphia   215-977-9758
                          215-977-9790#
                          215-977-9794*
  Pittsburgh   412-391-7732
                          412-391-8818
                          412-391-8218#
                          412-261-4192*
  Reading   215-375-4850
  York   717-845-7631
  Upper Darby   215-977-9758
                          215-977-9790#
                          215-977-9794*

Rhode Island
  Providence   401-823-8900

South Carolina
  Charleston   803-763-0090
                          803-556-0422#
  Columbia   803-783-5484
                          803-776-5355*
  Greenville   803-255-4686
  Myrtle Beach   803-238-8625

South Dakota
  Rapid City   605-341-3733

Tennessee
  Chattanonga   615-877-5804
  Gatlingburg   615-436-2001
  Knoxville   615-584-9902
  Memphis   901-452-1710
                          901-452-8530
                          901-452-2470#
                          901-323-0220*
  Nashville   615-366-1947
  Oak Ridge   615-483-2292

Texas
  Amarillo   806-379-8411
  Austin   512-444-7234
  Corpus Christi   512-887-2983
  Dallas   214-761-0599
                          214-761-9040
                          214-748-0976#
                          214-953-0436*
  El Paso   915-565-4670
                          915-565-4661
                          915-562-2617#
  Fort Worth   817-870-2468
                          817-870-2461+
  Houston   713-225-2330
                          713-225-2550+
                          713-225-2500#
                          713-225-0843*
  Lubbock   806-763-5081
  Midland   915-697-8211
  San Antonio   512-435-3883

Utah
  Provo   801-377-1120
  Salt Lake City   801-521-2890
                          801-521-2915
                          801-521-6326*

Vermont
  Burlington   802-862-1575

Virginia
  Alexandria   703-352-7500
                          703-841-9834
  Arlington   703-352-7500
                          703-841-9834
  Bethesda   703-352-7500
                          703-841-9834
  Charlottesville   804-973-8815+
  Chesapeake   804-461-6128
                          804-461-6167
  Fairfax   703-352-7500
                          703-352-8750#
                          703-352-3136
  Hampton   804-772-0016
  Midlothian   804-358-8774
  Norfolk   804-461-6167
                          804-461-6128
  Portsmouth   804-461-6128
                          804-461-6167
  Richmond   804-358-8274
  Roanoke   703-563-8421
  Virginia Beach   804-461-6128
                          804-461-6167

Washington
  Seattle   206-241-7023
                          206-241-9111
                          206-241-8137#
                          206-242-5767*
  Spokane   509-326-0515

West Virginia
  Charleston   304-768-9700
  Huntington   304-736-2331

Parkersburg   304-485-4225
Wheeling   304-233-9470

Wisconsin
  Brookfield   414-258-5616
  Madison   608-256-6525
  Milwaukee   414-258-5616
                     414-258-6049*

Wyoming
  Casper   307-234-6914

TERMINAL DESIGNATORS

Terminals supported by InfoPlex.

Terminal	Designator
AJ830	None required
Beta	U (auto line feed)
	B (no auto line feed)
CRT	C
Diablo	None required
Decwriter	None required
GE 300	GEIII
Hazeltine 2000	HAZMM
LA34	None required
LA36	None required
Memorex	M
NCR 260	N
Selectric	T
Teletype 33	T
Teletype 35	T
Terminet 300	G
Terminet 1200	GE
TI745	TI
TI765	TI
TI Silent 700	TI
Univac DCT 500	U (auto line feed)
	B (no auto line feed)

Dialnet

Arizona
Phoenix 602-257-8895

California
Long Beach 213-491-0803
Los Angeles 818-300-9000
Marina del Rey 213-305-9833
Newport Beach 714-756-1969
Oakland 415-633-7900
Palo Alto (300 Baud) 415-858-2575
  415-858-2461
Palo Alto (120 Bell) 415-858-0511
  415-858-2460
Palo Alto (120 VADIC) 415-858-2391
Palo Alto (120 B2020) 415-858-2421
Sacramento 916-444-5030
San Diego 619-297-8610
San Francisco 415-957-5910
Santa Clara 408-986-9610

Colorado
Denver 303-860-9800

Connecticut
Bloomfield 203-242-5954
Stamford 203-324-1201

Delaware
Wilmington 302-652-1706

District of Columbia

Washington 703-359-2500

Georgia
Atlanta 404-447-0374

Illinois
Chicago 312-341-1444

Indiana
Indianapolis 317-635-7259

Louisiana
Shreveport 318-688-5440

Maryland
Baltimore 301-234-0940

Massachusetts
Boston 617-524-2442
Lexington 617-862-6240

Michigan
Ann Arbor 313-622-4103
Detroit 313-964-1309
Kalamazoo 616-381-1603

Minnesota
Minneapolis 612-338-0676

Missouri
St. Louis 314-731-0122

New Jersey
Lyndhurst 201-438-5719
Morristown 201-292-9646
Newark 201-824-1412
Piscataway 201-562-9680
Princeton 609-683-5800
Trenton 609-888-0320

New York
Albany 518-458-9710
Buffalo 716-856-1022
Hempstead 516-489-6868
New York City 212-422-0410
Rochester 716-458-7300
White Plains 914-328-7810

North Carolina
Research Triangle Park 919-549-9290

Ohio
Cincinnati 513-489-3980
Cleveland 216-621-3807
Columbus 614-461-8348
Dayton 513-228-3470

Oklahoma
Oklahoma City 405-495-7620

Oregon
Portland 503-228-2771

Pennsylvania
Allentown 215-776-2030
Philadelphia 215-557-7711
Pittsburgh 412-471-1421
King of Prussia 215-768-0822

Texas
Dallas 214-631-9861
Houston 713-531-0505

Utah
Salt Lake City 801-532-3071

Virginia
Fairfax 703-359-2564

MCI Mail

Arizona	
Phoenix	602-266-1148
California	
Los Angeles	213-620-1449
Oakland	415-540-1114
Palo Alto	415-323-0251
Sacramento	916-442-6986
San Diego	619-268-1708
San Francisco	415-543-1560
San Jose	408-995-6711
Santa Ana	714-550-7128
Sherman Oaks	818-906-8989
Colorado	
Denver	303-831-8139
Connecticut	
Hartford	203-728-1909
Stamford	203-325-8133
District of Columbia	703-525-6500
Florida	
Clearwater	813-586-0955
Jacksonville	904-358-1749
Largo	813-586-0955
Miami	305-381-9012
St. Petersburg	813-586-0955
Tampa	813-221-1597
Georgia	
Atlanta	404-577-7363
Illinois	
Chicago	312-856-9000
Naperville	312-369-0805
Oakbrook	312-850-9511
Indiana	
Indianapolis	317-634-2208
Maryland	
Baltimore	301-583-6850
Massachusetts	
Boston	617-262-6468
Michigan	
Detroit	313-962-5980
Minnesota	
Minneapolis	612-893-9482
Missouri	
Kansas City	816-474-3169
St. Louis	314-991-1881
New Jersey	
Hackensack	201-488-2622
Newark	201-623-0295
New York	
Buffalo	716-847-6050
Garden City	516-596-0404
Long Island	516-596-0404
New York	212-245-0355
Rochester	716-248-8000
White Plains	914-232-7527
Ohio	
Cincinnati	513-651-1204
Cleveland	216-771-7177
Columbus	614-221-3451
Pennsylvania	
Philadelphia	215-636-9060
Pittsburgh	412-261-9918
Tennessee	
Memphis	901-523-9314
Texas	
Dallas	214-754-0461
Fort Worth	817-338-4159
Houston	713-850-1005
Washington	
Seattle	206-282-3077
Wisconsin	
Milwaukee	414-347-1769
WATS Access Number	800-323-0905

Datapac

Province/City	300 Baud	1200 Baud
Alberta		
Calgary	403-264-9340	403-290-0213
Edmonton	403-420-0185	403-423-4463
Fort McMurry	403-791-2884	403-743-5207
Grande Prairie	403-539-0100	403-539-6434
Lethbridge	403-329-8755	403-327-2004
Medicine Hat	403-526-6587	403-529-5521
Red Deer	403-343-7200	403-342-2208
British Columbia		
Kamloops	604-374-5941	604-374-9510
Kelowna	604-860-0331	604-860-9762
Nanaimo	604-753-6491	604-754-8291
Nelson	604-354-4411	604-354-4824
Prince George	604-564-4060	604-562-8469
Terrance	604-635-7221	604-562-8469
Vancouver	604-689-8601	604-684-7144
Victoria	604-388-9300	604-386-0900
Manitoba		
Brandon	204-725-0878	204-727-6609
Dauphin	204-638-9244	204-238-9441
Morden	204-822-6237	204-822-6287
Portg la Prairie	204-239-1166	204-239-1688
Selkirk	204-785-8625	204-785-8771
Steinbach	204-326-9826	204-326-1385
Thompson	204-778-4461	204-778-4451
Winnipeg	204-475-2740	204-943-4488
New Brunswick		
Bathurst	506-548-4461	506-548-4581
Campbellton	506-759-8561	506-759-8571
Edmondston	506-739-6621	506-739-6611
Fredericton	506-454-9462	506-454-4525
Moncton	506-854-7078	506-854-7510
New Castle	506-622-4451	506-622-8471
Saint John	506-693-7399	506-642-2231
Woodstock	506-328-9361	506-328-9351
Newfoundland		
St. John's	709-726-4920	709-726-5501
Nova Scotia		
Amherst	902-667-5035	902-667-5297
Bridgewater	902-543-6850	902-543-1360
Halifax	902-477-2000	902-477-8000
Kentville	902-678-1030	902-678-2096

New Glasgow	902-752-0944	902-752-1731
Sydney	902-539-7010	902-539-8040
Truro	902-662-3258	902-662-3773

Ontario
Barrie	705-737-4100	705-737-4120
Belleville	613-966-6002	613-966-9301
Brampton	416-791-8900	416-791-8950
Brantford	518-756-0000	416-756-0020
Brockville	613-345-0520	613-345-3780
Chalk River	613-589-2175	613-589-2117
Chatham	519-345-7710	519-354-7716
Clarkson	416-823-6000	
Cornwall	613-938-9700	613-938-4777
Galt	519-622-1714	519-622-1780
Guelph	519-836-7930	519-836-7960
Hamilton	416-523-6800	416-523-6900
Kingston	613-549-7720	613-549-7760
Kitchener-Waterloo	519-579-0009	519-579-0310
London	519-679-7500	519-679-7620
Niagara Falls	416-357-2702	416-375-2770
North Bay	705-476-3900	705-476-3920
Oshawa	416-579-8920	416-579-8950
Ottawa	613-567-9100	613-567-9300
Peterborough	705-748-6940	705-748-6945
St. Catharines	416-688-5620	416-688-5640
Sarnia	519-336-9920	519-336-0950
Sault Ste. Marie	705-942-4960	705-942-4970
Sudbury	705-673-9602	705-673-9652
Thunder Bay	807-623-9644	807-623-3270
Toronto	416-868-4000	416-868-4100
Windsor	519-973-1000	519-973-1020
Woodstock	519-485-5220	519-485-5236

Prince Edward Island
Charlottetown	902-566-5002	902-569-3784

Quebec
Drummondoille	819-477-7151	819-477-7153
Granby	514-375-1240	514-375-4184
Joliette	514-759-8340	514-759-8381
Jonquiere/Chicoutimi	418-545-2272	418-545-2290
Montreal	514-878-0450	514-878-0640
Quebec City	418-647-4690	418-647-2691
St. Hyacinthe	514-774-9270	514-774-9991
St. Jean	514-346-8779	514-347-6211
St. Jerome	514-432-3453	514-432-3165
Sherbrooke	819-655-2770	819-566-2990
Sorel	514-743-3381	514-743-0101
Trois Rivieres	819-373-2600	819-373-2603
Valleyfield	514-377-1260	514-377-1680

Saskatchewan
Moose Jaw	306-963-7611	306-374-5910
Prince Albert	306-922-4233	306-922-4234
Regina	306-565-0111	306-565-0181
Saskatoon	306-665-6660	306-665-7758
Swift Current	306-778-3941	306-778-3951

INDEX

Acoustic couplers 31
ACSP 36
Ambiset 17
Anderson-Jacobson 50
Apple II 12
Around 36
ASCII 27,35
ASCII PRO 37
Asher 51
AT&T Mail 59-61
AT&T Voicemail 17
Batch Services 11
Baud rates 333
Binary files 11
Bits 28
Blast 37
Cablegrams 162, 222, 252
Cermetek Modem 53-54
Commodore 12, 64
Communication software 28-30
COMMX PAC 38
Compaq 12, 15-16
CompuServe access numbers 272
CompuServe, see EasyPlex and InfoPlex
Computer letters 163
Computer telephone 15
Computer to voice mail 59-61
Computers 13-15, 27-28
Cosystem 16
Courier 2400 51
CrossTalk 39
CrossTalk-XVI 39
Datapac access numbers 280
Delphi
   Background 93
   Command 96
   Conclusion 108
   Connection procedures 95
   Downlading files 101
   Files 103
   Folders 103
   Help 97
   Hours 94
   Logoff procedures 108
   Logon procedures 95
   Mailing lists 101-102
   Network access 94
   Prompts 96
   Rates 99
   Reading mail 100-103
   Sending letters 98-100
   Storage 94
   Subscription 93-94
   Telex 107
   Translation services 107
   Uploading files 100-101
   Using the service 96, 98
Dialcom, see ITT Dialcom
Dialmail
   Background 119
   Conclusion 130
   Connection procedures 120
   Help 122
   Hours 120
   Logoff procedures 130
   Logon procedures 121
   Mailing lists 125
   Network access 120
   Paper mail 129-130
   Prompts 122
   Rates 119
   Reading mail 127-129
   Scanning mail 125-127
   Sending letters 122
   Storage 120
   Subscription 119
   Using the service 121-122
Dialnet access numbers 278
EasyLink
   Background 148
   Cablegrams 162
   Commands 151
   Computer letters 163
   Conclusion 165
   Connection procedures 150
   Express documents 164

Help 152
Hours 149
Logoff procedures 165
Logon procedures 151
Mailgrams 160
Mailing lists 156
Network access 150
Rates 119
Reading mail 127-129
Scanning mail 125-127
Sending letters 122
Storage 120
Subscription 119
Using the service 121-122
EasyPlex
   Address book 88
   Background 80
   Commands 83
   Conclusion 92
   Connection procedures 82
   Help 84
   Hours 81
   Logoff procedures 92
   Logon procedures 83
   Network access 82
   Prompts 83
   Rates 81
   Reading mail 89
   Sending messages to MCI Mail 91
   Shortcuts 90
   Storage 82
   Subscription 81
   Uploading files 87
   Using the service 84-85
ECHO
   Background 109
   Conclusion 118
   Files 116
   Folders 116
   Help 111
   Hours 110
   Logoff procedures 118
   Logon procedures 111
   Mailing lists 115
   Network access 110
   Paper mail 117
   Rates 110
   Reading mail 116
   Sending letters 112-114
   Storage 110, 112

   Subscription 110
   Uploading files 114
   Using the service 111
Edit 34
Epson 15
Equipment requirements 12
ERA 2 52
Error-checking 11
ESTeem 21
Express delivery service 8
Express documents 164
Facsimile transmitter 8
FM SCA Receivers 19-21
FM Subcarriers 19-20
Forms 195
Framework II 39-40
GENIE 232
Grid 15
Hardware 12
Hazeltine 12
Hewlett-Packard 15
IBM PC 12, 15-16
IBM 5841/PC internal modem 52
InfoPlex
   Background 166
   Conclusion 178
   Connection procedures 167-168
   Help 169-170
   Logoff procedures 177-178
   Logon procedures 168-169
   Network access 167
   Reading mail 174-176
   Sending letters 170-173
   Subscription 166
   Telex 176-177
   Uploading files 173-174
   Using the service 169
InfoPlex terminal designators 277
International Telex 222
ITT Dialcom
   Background 233
   Cablegrams 251
   Conclusion 252
   Connection procedures 234
   Files 246
   Help 236
   Holding mail 247
   Lettergrams 251-252
   Logoff procedures 252
   Logon procedures 235-236

Mailgrams 251
Network access 234
Rates 2345
Reading mail 244-246
Shortcuts 247-248
Speedmail 250
Storage 234
Subscription 234
Uploading files 242-244
Using the service 236-237
ITT Modem 54
Kaypro 12, 15
Knowledge Index, see Dialmail
Lettergram 251-252
LYNC 44
Macintosh 12
Mail-Com 40-41
Mailgrams 78, 160, 222, 251
Mastercom 41
Maxwell 1200V/1200PC 52-53
MCI Mail
   Advanced service 132
   CompuServe messages
   Conclusion 147
   Connection procedures
   Editing mail 138-139
   Help 135
   Logoff procedures 147
   Logon procedures 134
   Mail alert 133, 140
   Mailing lists 141
   Network access 133
   Paper mail 132
   Rates 132
   Reading mail 143
   Sending letters 135-137
   Shortcuts 145
   Subscription 132
   Telex 18, 133, 145, 146
   Uploading files 142
   Using the service 134
MCI Mail access numbers 279
Micro-Link II 41-42
Microphone 42
Microsoft Access 43
Mite 43-44
Modem reviews 50-57
Modems 30-33, 50-57
Move-It 44-45
NEC 15

Network access numbers 253-281
OnTyme
   Background 210
   Conclusion 223
   Connection procedures 211
   Files 220-222
   Folder 220
   Help 213
   Hours 210
   Logoff procedures 222-223
   Logon procedures 211-212
   Mailing lists 217
   Network access 211
   Prompts 212
   Rates 210
   Reading mail 218-220
   Sending letters 214-216
   Subscription 210
   Uploading files 216-217
   Using the service 212
Overseas priority letters 164
P/C Privacy 45
Paper mail 9, 117, 120-121, 163-165, 231, 250-251
PC Write 34
PC-Talk III 45-46
Perfect Link 46
PFS Access 46-47
Pony express 4, 7
Popcom X100 53
Portable computers 13-15
Postal service 6, 7-8, 25-26
PostPlus 47-48
Qmodem 48
GE Quik-Comm
   Background 224
   Conclusion 232
   Connection procedures 226
   Genie 232
   Help 227
   Logoff procedures 232
   Logon procedures 226
   Network access 225
   Paper mail 231
   Quik-Grams 231
   Rates 225
   Reading mail 229
   Sending letters 227-229
   Software 224
   Subscription 225

Telex 231-232
  Using the service 226
Quik-Gram 231
Qume 12
RCA MAIL
  Background 199
  Conclusion 209
  Connection procedures 201
  Hours 200
  Logoff procedures 208
  Logon procedures 201
  Network access 201
    Rates 200
  Reading mail, see Telemail
  Scanning mail, see Telemail
  Sending letters, see Telemail
  Shortcuts
  Storage 200
  Subscription 200
  Telegrams 207-208
  Telex 199, 202-208
  Using the service 201-202
Reachout 48-49
Relay Gold 50
RS-232 27
Satellite broadcast 19,21
Security modems 53-54
Serial interface see RS-232
Sharp 15
Smart Cat 55
SmartCom II 49
SmarTeam 55
Smartmodem 1200/2400 54-55
Softwre reviews 36-50
Sord 15
The Source Service
  Background 65
  Conclusion 79
  Connection procedures 65-66
  Files 73
  Folders 74
  Help 66-67
  Hours 64
  Logon procedures 66
  Mailgrams 78
  Mailing lists 71
  Network access 64-65
  Rates 64
  Reading mail 74-76
  Sending letters 67-71
  Shortcuts 76
  Storage 64
  Subscription 63-64
  Uploading files 72
  Using the service 66
TeleCompaq 15-16
Telecomputer 12
Telegrams 161, 222
GTE Telemail
  Background 179
  Conclusion 198
  Connection procedures 181
  Help 182
  Hours 180
    Inform Script 195
  Logoff procedures 198
  Logon procedures 181
  Mailing lists 186-187
  Network access 180
  Rates 180
  Reading mail 188-190
  Sending letters 183-185
  Shortcuts 194
  Storage 180
  Subscription 180
  TelemailXpress 179
    Telex 195-198
  Trainer 179
  Using the service 181
Telenet access numbers 261
Telenet terminal identifiers 266
Telex 17-19, 107, 133, 145-147, 159-160, 176-177,195-198, 204-208, 222, 231-232
Terminal 26
Texas Instruments 15
Text editors 11
TI 99/4A 12
Translation services 107
Tymnet access numbers 254
UAP-LINK 49
Uninet access numbers 268
Unix PC 16
Uploading, preparing text for 35
VA 212 55
VBI Signals 19-20
VN Relay 50
Voice mail 17, 58-62
Watson 56-57
Western Union Telex 18
Wireless communications 21-22

Word processing 34-35
WorldWide Telex 148
Xerox 12
Zaisan ES.3 16
Zenith 15
ZOOM Modem PC 1200 57

TRADEMARKS AND SERVICEMARKS

ABI/INFORM is a trademark of Data Courier, Inc. ACSP IBM is a trademark of International Business Machines Corporation. AJ 347 is a trademark of Anderson-Jacobson, Inc. AJ CONNECTION is a trademark of Anderson-Jacobson Inc. AJ 1212-AD1 is a trademark of Anderson-Jacobson, Inc. APPLE II is a trademark of Apple Computer, Inc AROUND is a trademark of Teal Communications. ASCII PRO is a trademark of United Software Industries. ASHER is a trademark of Asher Technologies, Inc. BLAST is a trademark of the Communication Research Group. CERMETEK is a trademark of Cermetek Microelectronics. COMMX-PAC is a trademark of Hawkeye Grafix. COMPAQ is a trademark of Compaq Computers. COMPUSERVE, INFOPLEX and COMPUSERVE INTERCHANGE are trademarks of CompuServe Information Services. COSYSTEM is a trademark of Cygnet Communications. COURIER is a trademark of US Robotics CP/M and CP/M-86 are trademarks of Digital Research, Inc. CROSSTALK is a trademark of Microstuf, Inc. DATAPAC is trademark of the Commonwealth of Canada. DELPHI is a trademark of General Videotex Corporation. DIALCOM is servicemark of ITT Dialcom, Inc. DIALOG is a trademark of Dialog Information Services, Inc. DIALMAIL is a trademark of Dialog Information Services, Inc. DIALNET is a servicemark of Dialog Information Services, Inc. DOW JONES NEWS/RETRIEVAL is a trademark of Dow Jones & Company. EASYLINK is a servicemark of Western Union Telegraph Company. EASYPLEX is a trademark of CompuServe Information Services. ECHO is a trademark of the Electronic Communications for the Home and Office. EDIT is a trademark of PC-SIG. ERA 2 is a trademark of Microcom, Inc. GRAPHNET is a trademark of Graphnet Systems, Inc. IBM and IBM PC are trademarks of International Business Machines Corporation. INSTANTMAIL is a servicemark of Western Union Telegraph Company. LYNC is a trademark of Norton-Lambert Corporation. KNOWLEDGE INDEX is a servicemark of Dialog Information Services, Inc. LEARN is servicemark of ITT Dialcom, Inc. MACINTOSH is a trademark of Apple Computer, Inc. MAILGRAM is a trademark of Western Union Telegraph Company. MARK III is a trademark of General Eclectic Information System Company MASTERCOM is a trademark of Software Store, Ltd. MAXWELL, GEORGE, & VA 212 are trademarks of Racal-Vadic MCI MAIL, MCI LETTER, INSTANT LETTER, OVERNIGHT LETTER, and FOUR HOUR LETTER are servicemarks of MCI Communications Corporation. MICROSOFT ACCESS is a trademark of Microsoft Corporation. MICRO LINK II is trademark of Digital Marketing. MITE is a trademark of Mycroft Labs, Inc. MOVE-IT is a trademark of Woolf Software. MS DOS is a trademark of Microsoft. NEWS-TAB is a servicemark of ITT Dialcom. OAG is a trademark of Dun & Bradstreet PCMAIL is a servicemark of ITT Dialcom, Inc. PC MAILBOX is a trademark of General Electric Information System Company. PC TALK III is a trademark of The Headlands Press. PC WRITE is a trademark of Quick Soft. PFS:ACCESS is a trademark of Software Publishing Corporation. PERFECT LINK is a trademark of Thorn-EMI Computer Software. POS PLUS is a trademark of MCTel, Inc. QMODEM is a trademark of The Forbin Project, Inc. QUIK-COMM & QUIK-GRAM are trademarks of General Electric Information System Company. RADIO SHACK is a trademark of Tandy Corporation. RC MAIL is a trademark of RCA Global Communications. REACHOUT is a trademark of Applied Computer Technique RELAY GOLD and VM RELAY are a trademarks of VM Personal Computing. SECURITY MODEM is a trademark of IT Data Equipment & Systems. SMARTCOM II and SMARTMODEM are trademarks of Hayes Microcomputer Products SMARTEAM is a trademark of Morrison & Dempsey Communications. SMART-CAT is a trademark of Novation. THE SOURCE is a resigtered servicemark of Source Telecomputing Corporation. TELECOMPAQ is a trademark of Compaq Computers. TELEGRAM and TELEX are trademarks of Western Union Telegraph Company. TELEMAIL is a resigtered servicemark of GTE Telenet Communications. TYMNET is a trademark of Tymshare, Inc. UAP-LINK is a trademark of Unique Automation Products. UNINET is a trademark of United Telecom Communications, Inc. UNIX PC is a trademark of AT&T. VOLKSMODEM is a trademark of Anchor Automation. WATSON is a trademark of Natural Microsystems Corporation. WORDSTAR is a trademark of MicroPro International. XEROX is a trademark of Xerox Corporation. XMA is a servicemark of ITT Dialcom, Inc.

Also available from Steve Davis Publishing

The Facts On
FAX

The Manager's Guide to Facsimile Communications Today

By Lawrence Robinson

In today's business climate, managers need instant information to make timely, profitable decisions. Today's new generation of facsimile terminals can provide any business, large or small, with quick, cost-effective communications. Facsimile (or "fax" for short) is an effective and affordable form of "electronic mail," allowing any business to transmit a page of information, even graphics, anywhere in the world in less than a minute.

This book shows you how today's advanced fax machines can help you increase productivity and profits while reducing costs in your business. Here is all the information you need to select and use modern fax terminals to improve the efficiency of your company's communications, including a guide to examining your communications needs and tips on setting up your own fax network. A bonus "Buyer's Guide" section highlights many of the top machines on the market for all sizes of businesses. ISBN 0-911061-15-0. $19.95 US.

SPECIAL FREE BONUS!

Exclusively for readers of *The Electric Mailbox*

Discounts worth over $200

To help you get started in the wonderful world of computerized communications, we have arranged with several major vendors of related products and services to provide you with special introductory offers and discounts on:

- Electronic mail services.
- Online database services.
- Communications software.
- Modems.
- Computer books, and more.

To receive your bonus pack of special discount offers worth over $200, just complete this form and send it along with $1.00 to cover postage and handling. Please print clearly.

Send to: STEVE DAVIS PUBLISHING
P.O. Box 190831
Dallas, Texas 75219.

Name _____

Address _____

City _____ State _____ Zip _____

Store where you purchased this book: _____